Metal and Non-Metal Biguanide Complexes

R.K. Ray
M.Sc., Ph.D.
Former Head
Department of Chemistry
Rama Krishna Mission Vivekananda
Centenary College, Rahara
West Bengal (India)

G.B. Kauffman
Ph.D. (University of Florida)
Professor, Department of Chemistry
California State University
Fresno, USA

PUBLISHING FOR ONE WORLD

NEW AGE INTERNATIONAL (P) LIMITED, PUBLISHERS
New Delhi • Bangalore • Chennai • Cochin • Guwahati • Hyderabad
Jalandhar • Kolkata • Lucknow • Mumbai • Ranchi
Visit us at **www.newagepublishers.com**

Published by New Age International (P) Ltd., Publishers
First Edition: 1999 (PB)
First Edition: 2010 (HB)

Branches:

- No. 37/10, 8th Cross (Near Hanuman Temple), Azad Nagar, Chamrajpet, **Bangalore**-560 018. Tel.: (080) 26756823, Telefax: 26756820, E-mail: bangalore@newagepublishers.com
- 26, Damodaran Street, T. Nagar, **Chennai**-600 017. Tel.: (044) 24353401, Telefax: 24351463 E-mail: chennai@newagepublishers.com
- CC-39/1016, Carrier Station Road, Ernakulam South, **Cochin**-682 016. Tel.: (0484) 2377004, Telefax: 4051303. E-mail: cochin@newagepublishers.com
- Hemsen Complex, Mohd. Shah Road, Paltan Bazar, Near Starline Hotel, **Guwahati**-781 008. Tel.: (0361) 2513881. Telefax: 2543669, E-mail: guwahati@newagepublishers.com
- No. 105, 1st Floor, Madhiray Kaveri Tower, 3-2-19, Azam Jahi Road, Nimboliadda, **Hyderabad**-500 027. Tel.: (040) 24652456, Telefax: 24652457, E-mail:hyderabad@newagepublishers.com
- RDB Chambers (Formerly Lotus Cinema)106A, 1st Floor, S.N. Banerjee Road, **Kolkata**-700 014. Tel.: (033) 22273773, Telefax: 22275247, E-mail:kolkata@newagepublishers.com
- 16-A, Jopling Road, **Lucknow**-226 001. Tel.: (0522) 2209578, 4045297, Telefax: 2204098 E-mail: lucknow@newagepublishers.com
- 142C, Victor House, Ground Floor, N.M. Joshi Marg, Lower Parel, **Mumbai**-400 013. Tel.: (022) 24927869. Telefax: 24915415, E-mail: mumbai@newagepublishers.com
- 22, Golden House, Daryaganj, **New Delhi**-110 002. Tel.: (011) 23262370, 23262368, Telefax: 43551305. E-mail: sales@newagepublishers.com

ISBN : 978-81-224-2749-3

Rs. 495.00

C-09-06-3739

Printed in India at Glorious Printers, Delhi.

PUBLISHING FOR ONE WORLD
NEW AGE INTERNATIONAL (P) LIMITED, PUBLISHERS
4835/24, Ansari Road, Daryaganj, New Delhi-110002
Visit us at **www.newagepublishers.com**

Dedicated to
Professor Rabindra Lal Dutta
who introduced and guided R.K. Ray
through many puzzling but challenging
areas of chemical research, with faith and reverence

The chemists are a strange class of mortals impelled by an almost insane impulse to seek their pleasure among smoke and vapour, soot and flame, poisons and poverty; yet among all these evils I seem to live so sweetly that I may die if I would change places with the Persian king.

—Johann Joachim Becher

Experiments are the only means of knowledge at our disposal. The rest is poetry, imagination.

—Max Planck

FOREWORD

Even as a young graduate student working under Professor John C. Bailar, Jr. at the University of Illinois during the early 1940s, I was impressed by the multi-faceted contributions of Indian researchers to coordination chemistry. Being interested in the reaction mechanisms of coordination compounds, I can vividly recall the Rây-Dutt rhombic twist mechanism proposed at that time to account for racemizations and other intramolecular rearrangements of optical isomers that do not involve dissociation (P. Rây and N.K. Dutt, *J. Indian Chem. Soc.,* 18(1941) 289; 20(1943) 81). Another area of research in which Indian chemists have been active has been the stabilisation of unusual oxidation states, e.g., silver(II) and silver(III), by complexation (D. Sen, P. Rây, and N.N. Ghosh, *J. Indian Chem. Soc.*, 27(1950) 619; P. Rây and D. Sen, Chemistry of Bi- and Tripositive Silver, National Institute of Sciences of India, New Delhi, (1960). And, of course, it is to Indian chemists that we owe much of our fundamental knowledge of complexes with organic nitrogen donor ligands.

Among the most extensively studied of such ligands is biguanide and its derivatives, "because of their powerful coordinating ability, medicinal value, and the wide variety of their complexes, which have elucidated problems of valency, structure, stability and other fundamental aspects of coordinating chemistry", as Drs. Ray and Kauffman so aptly express it. In 1961, a classic review of metal complexes of biguanide and guanylureas and their derivatives by the late dean and elder statesman of Indian coordination chemistry, Priyadaranjan Rây, appeared in *Chemical Reviews*. Since then, research on these complexes has increased greatly, and the need for a comprehensive review, updated to the present time, has become a *sine qua non* for researchers in the field and others interested in the unusual properties of these compounds.

In this short but authoritative book, Drs. Ray and Kauffman succinctly review all important studies on complexes of biguanide and their derivatives published since Priyadranjan Rây's classic chemical Reviews article. By giving special attention to synthesis, properties, magnetism, spectroscopy, reactions, kinetics, structures and applications, this book is the only one of its kind. Structures of the ligands and their complexes as revealed by magnetic moments, electronic spectra, EPR and X-ray crystallography have been considered. The book's references which number more than four hundred should give practising scientists ready access to the crucial details needed for research and development on these theoretically and practically significant compounds. As such, it provides data that are widely scattered is in scientific literature in a single handy source book. I heartily recommend it to inorganic, organic, medical and pharmaceutical chemists, as well as to anyone concerned with this unique class of ligands and their complexes with a wide variety of elements of varying oxidation states and coordination numbers.

FRED BASOLO
Charles E. & Emma H. Morrison
Professor of Chemistry
Northwestern University
Evanston, Illinois

PREFACE

Among nitrogen donor ligands biguanides are unique and have been extensively investigated because of their powerful coordinating ability, medicinal applications and the wide variety of their coordination compounds, which have thrown light on problems of valency, structure, stability and other fundamental aspects of coordination chemistry. They have also found use as analytical reagents, as germicidal, bacteriostatic, antihypoglycaemic and anticarcinogenic agents, and as medicines for the treatment of malaria, filaria and influenza. They are starting materials for numerous triazines and medically useful heterocyclic compounds. In the free state or in the form of their complexes, they undergo Schiff base-type reactions to yield a variety of products. They can stabilize a number of unusual oxidation states of the transition metals to which they coordinate, and they yield coordination compounds of great stability.

Since the appearance of Priyadaranjan Rây's review of the metal complexes of biguanide and its derivatives (*Chem. Rev.*, 61(1961) 313), the pace of research on these compounds have accelerated tremendously. Therefore, a review updating this topic must deal with an extremely large number of published studies. In this book we have tried to include all important articles published since Ray's review through 1992 as well as a number of later studies with special emphasis on synthesis, magnetism, spectroscopy, reaction kinetics and structure. These studies have greatly contributed to the richness and complexity of modern experimental and theoretical coordination chemistry. We hope that our book will not only be valuable to workers in the field but will also encourage others to enter an exciting field of research that is replete with theoretical problems and practical applications.

R.K. Ray
G.B. Kauffman

ACKNOWLEDGMENTS

The authors thank Mary McIntyre, Department of Chemistry, California State University, Fresno, for technical assistance and the following for permission to reproduce figures : The Royal Society of Chemistry, *J. Chem. Soc., A* (Figs. 20, 21, 22, 24), *J. Chem. Soc., Chem. Communications* (Fig. 43), Acta crystallographica (Figs. 40, 41, 45, 47, 48, 51, 52), *Nature* (Figs. 49, 50), Transition Metal Chemistry (Figs. 42, 79, 80, 81), *Indian Journal of Chemistry* (Fig. 56) Polyhedron (Figs. 63, 64), *Journal of Indian Chemical Society* (Figs. 114, 115). R.K. Ray wishes to thank Prof. H. Yoneda, Hiroshima University, Japan, Prof. D. H. Busch, University of Kansas, U.S.A., Dr. Peter Hubberstey, University of Nottingham, England, for providing reprints as well as facilities, and Prof. B.N. Mukherjee, Department of Chemistry, Dr. Dwijesh Majumder and Prof. M.K. Bandyopadhyay, Department of Physics, R.K. Mission V.C. College, Rahara, India, for encouragement. He is also indebted for the encouragement and patient services of Smt. Anushila Ray, Subhajit Ray, and Shankha Ray in preparation of this book are gratefully appreciated. G.B.K. wishes to thank his wife Lauria for her encouragement during the composition of this book.

ABBREVIATIONS

AAUH	1-amidino-*O*-alkylurea
AA^{i}UH	1-amidino-*O*-isoamylurea
AA^{n}UH	1-amidino-*O*-*n*-amylurea
AB^{i}UH	1-amidino-*O*-isobutylurea
AB^{n}UH	1-amidino-*O*-*n*-butylurea
acac	Deprotonated anion of acetylacetone
acacH	Acetylacetone
acan	Deprotonated anion of acetoacetanilide
acanH	Acetoacetanilide
AcH	Acetic acid
AEBG	2-aminoethyl biguanide
AEUH	1-amidino-*O*-ethylurea
AUH	amidinourea
α-alan$\overline{O}$	Deprotonated anion of α-alanine
α-alanOH	α-alanine
β-alan$\overline{O}$	Deprotonated anion of β-alanine
β-alanOH	β-alanine
AMUH	1-amidino-*O*-methylurea
AP^{i}UH	1-amidino-*O*-isopropylurea
AP^{n}UH	1-amidino-*O*-n-propylurea
$AspH_2$	Aspartic acid
benzyl bigH	N^1—benzyl biguanide
bigH	Biguanide
$bigH_2^+$	Mono protonated biguanide
$bigH_3^{2+}$	Diprotonated biguanide
biH_2	Biuret
B.M.	Bohr magneton
bpha	Deprotonated anion of N-benzoyl phenylhydroxylamine
bphaH	N-benzoyl phenylhydroxylamine
But^n–bigH	N^1-*n*-butylbiguanide
C.D	Circular dichroism
ChlorobigH	Chloro biguanide
CSH	*d*-Camphor sulphonic acid

cys	Cysteine
d and *l*	Observed optical rotation (*dextro* and *levo*)
dichlorobigH	Dichloro biguanide
dichloro Me-bigH	Dichloro methyl biguanide
dipic	Deprotonated anion of dipicolinic acid (pyridine-2, 6-dicarboxylic acid)
dipicH_2	Dipicolinic acid (Pyridine-2, 6-dicarboxylic acid)
dipy	α, α′-dipyridyl
d.m.e	dropping mercury electrode
DMF	*N*, *N*-dimethylformamide
DMG	Deprotonated anion of dimethylglyoxime
$DMGH_2$	Dimethylglyoxime
DMSO	Dimethyl sulphoxide
DTA	Differential thermal analysis
EDTA	Deprotonated anion of ethylenediamine tetraacetic acid
en	Ethylenediamine
en$(bigH)_2$	N^1, N^1-ethylenedibiguanide
en$(bigH_2^+)_2$	Protonated N^1, N^1-ethylenedibiguanide
EPR	Electron paramagnetic resonance
ESR	Electron spin resonance
Et-bigH	N^1-ethyl biguanide
Et-bigH_2^+	Protonated N^1-ethyl biguanide
Et_2-bigH	N^1, N^1-diethyl biguanide
Et_2-bigH_2	Protonated N^1, N^1-diethyl biguanide
Et-NH_2	Ethylamine
Et_2-NH_2	Diethylamine
EtOH	Ethanol
gly$\overline{O}$	Deprotonated anion of glycine
glyOH	Glycine
hexbigH	N^1-hexyl biguanide
hexbigH_2^+	Protonated N^1-hexyl biguanide
hex$(bigH)_2$	N^1, N^1-hexamethylene dibiguanide
hex$(bigH_2^+)_2$	Protonated N^1, N^1-hexamethylene dibiguanide
hist$\overline{O}$	Deprotonated anion of histidine
histOH	Histidine
IDA	Deprotonated anion of iminodiacetic acid
$IDAH_2$	Iminodiacetic acid
IR	Infrared
leucO	Deprotonaled anion of leucine
leucOH	Leucine
Lut	Deprotonated anion of lutidinic acid (Pyridine-2, 6-dicarboxylic acid)

$LutH_2$	Lutidinic acid (Pyridine-2, 6-dicarboxylic acid)
Me-bigH	N^1-methyl biguanide
Me-$bigH_2^+$	Protonated N^1-methyl biguanide
Me_2-bigH	N^1, N^1-dimethyl biguanide
Me_2-$bigH_2^+$	Protonated N^1, N^1-dimethyl biguanide
Me-NH_2	Methylamine
Me_2-MH_2	Dimethylamine
MeOH	Methanol
m.p.	Melting point
m-phen$(bigH)_2$	N^1, N^1-m-phenylene dibiguanide
m-phen$(bigH_2^+)_2$	Protonated N^1, N^1-m-phenylene dibiguanide
methiO	Deprotonated anion of methionine
methioH	Methionine
morph-bigH	Morpholine biguanide
morph-$bigH_2^+$	Protonated morpholine biguanide
naph-bigH	N^1-naphthyl biguanide
naph-$bigH_2^+$	Protonated N^1-naphthyl biguanide
nitrophen	5 nitro-*o*-phenan throline
NMR	nuclear magnetic resonance
O-phen	Orthophenanthroline
Ox	Deprotonated anion of oxalic acid
OxH_2	Oxalic acid
Oxin	Deprotonated anion of 8-hydroxy quinoline
OxinH	8-hydroxy quinoline
O-Cl-Ph-bigH	O-Chlorophenylbiguanide
PaludH	Paludrine or [1-(p-Chlorophenyl)-5-isoprophyl biguanide
$PaludH_2^+$	Protonated paludrine or protonated [1-(p-cholorophenyl) -5-isopropyl biguanide]
Pbsa	Deprotonated anion of 1-phenyl biguanide-p-sulphonic acid
PbsaH	1-phenyl biguanide-p-sulphonic acid
PBz	2 – (2′-pyridyl) benzimidazole
Ph-alanOH	Phenyl alanine
Ph-bigH	N^1-phenyl biguanide
Ph-$bigH_2^+$	Protonated N^1-phenyl biguanide
Cl-Ph-bigH	N^1-Chlorophenyl biguanide
Cl-Ph-$bigH_2^+$	N^1-Protonated chlorophenyl biguanide
Ph (Et)-bigH	N^1, N^1-Phenylethyl biguanide
Ph (Et)-$bigH_2^+$	N^1, N^1-Protonated phenylethyl biguanide
PMR	Proton magnetic resonance

$Ph\text{-}NH_2$	Aniline
Pr^i-bigH	N^1-isopropylbiguanide
Pr^i-$bigH_2^+$	Protonated N^1-isopropylbiguanide
Pr^n-bigH	N^1-n-propylbiguanide
Pr^n-$bigH_2^+$	Protonated N^1-n-propylbiguanide
$pipz(bigH)_2$	Piperazine dibiguanide
$pipz(bigH_2^+)_2$	Protonated piperazine dibiguanide
Pn	Propylenediamine
picam	2-picolylamine
pic	Deprotonate anion of picolinic acid
picH	Picolinic acid
p-tol-bigH	N^1-(p-tolyl)biguanide
p-tol-$bigH_2^+$	Protonated N^1-(p-tolyl)biguanide
py	Pyridine
quinal	Deprotonated anion of quinaldenic acid
quinalH	Quinaldinic acid
R-bigH	N^1-alkyl biguanide
R-$bigH_2^+$	Protonated N^1-alkyl biguanide
sar	Deprotonated anion of sarcosine
sarH	Sarcosine
tart	Tartrate
T.G.	Thermogravimetry
trien	Trimethylenediamine
$tetme(bigH)_2$	Tetramethylenedibiguanide
$tetme(bigH_2^+)_2$	Protonated tetramethylenedibiguanide
$trien(bigH)_2$	Triethylenedibiguanide
$trime(bigH)_2$	Trimethylene dibiguanide
$trime(bigH_2^+)_2$	Protonated trimethylenedibiguanide
tolyl bigH	N^1-tolyl biguanide
tolyl $bigH_2^+$	Protonated N^1-tolylbiguanide
UV	Ultraviolet
$val\bar{O}$	Deprotonated anion of valine
valOH	Valine
XPS	X-ray photoelectron spectroscopy

CONTENTS

1

INTRODUCTION

Chemically and structurally, biguanide (diguanide or guanylguanidine) (I) (Fig. 1), 1-amidino-*O*-alkylurea (guanylalkylurea) (II) (Fig. 2), and guanylurea (dicyandiamidine or amidinourea) (III) (Fig. 3), and biuret(IV) (Fig. 4) are closely related compounds:

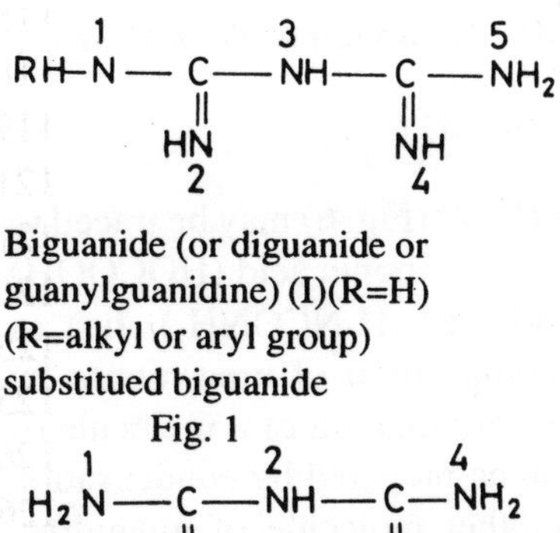

Biguanide (or diguanide or guanylguanidine) (I)(R=H) (R=alkyl or aryl group) substitued biguanide

Fig. 1

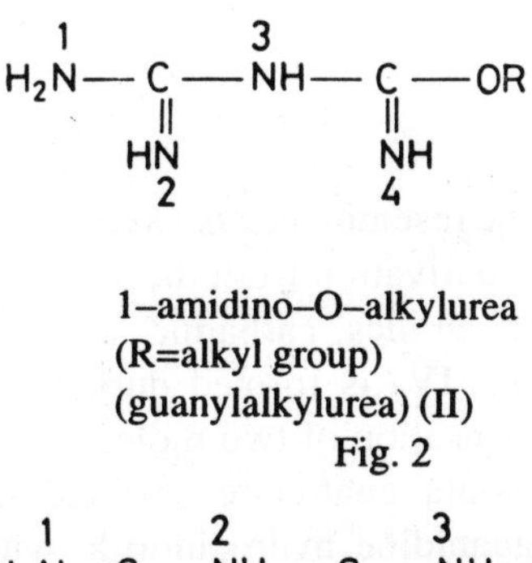

1–amidino–O–alkylurea (R=alkyl group) (guanylalkylurea) (II)

Fig. 2

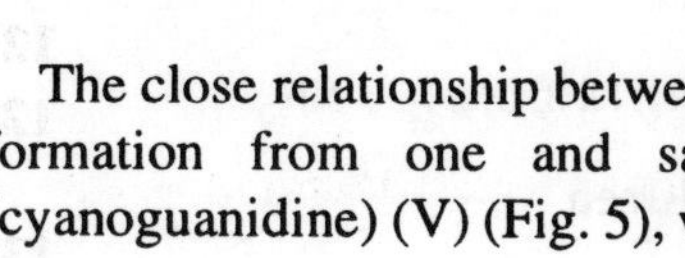

guanylurea (dicyandiamidine or amidinourea) (III)

Fig. 3

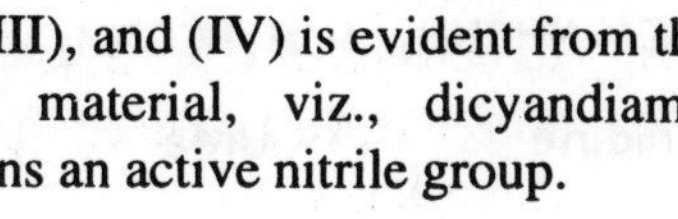

Biuret (IV)

Fig. 4

The close relationship between (I), (II), (III), and (IV) is evident from their formation from one and same source material, viz., dicyandiamide (cyanoguanidine) (V) (Fig. 5), which contains an active nitrile group.

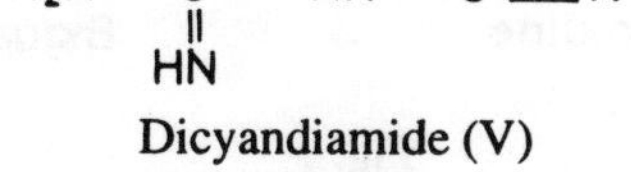

Dicyandiamide (V)

Fig. 5

In the presence of water and acid, dicyandiamide (V) takes up a molecule of water and forms guanylurea, also known as amidinourea or dicyandiamidine (III), while in the presence of ammonia and amines, a powerful coordinating ligand, biguanide or a substituted biguanide (I), is formed [1]. Metal-promoted addition of alcohol to dicyandiamide favours the formation of l-amidino-*O*-alkylureas (II) [2-10]:

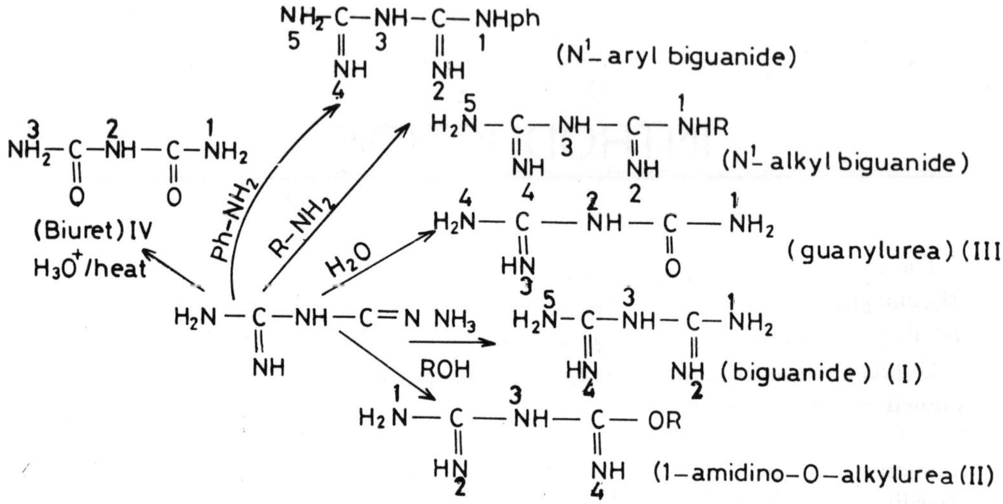

Fig. 6

The resemblance between the above ligands (I-IV) (Fig. 6) may be traced to their derivation from the same parent source, viz., carbonic acid (HOCOOH) or its amides, carbamic acid (H_2NCOOH) and urea (H_2NCONH_2). Just as biuret (IV) is formed during the thermal decomposition of urea nitrate by condensation of two molecules of urea with the elimination of a molecule of ammonia, guanylurea (II) and biguanide (I) can be prepared by condensation of guanidine hydrochloride with urea and another molecule of guanidine, respectively [11, 12] (Fig. 7):

$$H_2NCONH_2 + H_2NCONH_2 \longrightarrow H_2CONHCONH_2 + NH_3$$

Urea Urea Biuret

$$H_2NC(=NH)NH_2 + H_2NCONH_2 \longrightarrow H_2NC(=NH)NHCONH_2 + NH_3$$

Guanidine Urea Guanylurea

$$H_2NC(=NH)NH_2 + H_2NC(=NH)NH_2 \longrightarrow H_2NC(=NH)NHC(=NH)NH_2 + NH_3$$

Guanidine Guanidine Biguanide

Fig. 7

Of the four closely related compounds, biuret (IV) forms anionic complexes with transition metal ions such as Ley and Werner's [13] well known violet copper (II) biuret complex (VI) (Fig. 8):

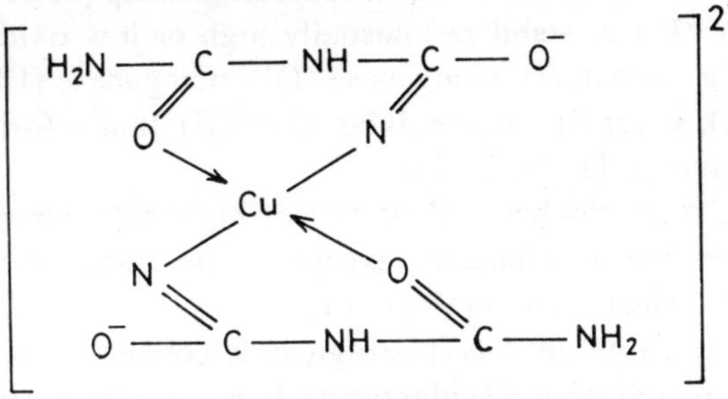

Copper(II) Biuret Complex (VI)
Fig. 8

The remaining three ligands, viz., biguanides (I), 1-amidino-*O*-alkylureas (II), and guanylurea (III), form well defined cationic complexes with transition metals [1, 5, 14].

Of these four nitrogen donor ligands, biguanides (I) have been extensively studied because of their powerful coordinating ability [1, 14], medicinal value [15, 16], and the wide variety of their complexes, which have elucidated problems of valency, structure, stability, and other fundamental aspects of coordination chemistry. More specifically:

1. Biguanide (I) behaves as a base, acid, zwitter ion (Fig. 9) or polar molecule [1].

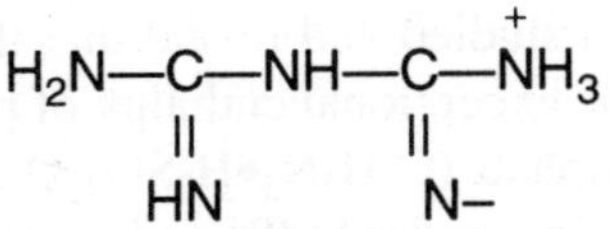

Fig. 9

2. Biguanide (I) has a high proton affinity in aqueous solution, its basic strength ($pK_1 = 13.0$) being only slightly lower than that of the hydroxyl ion ($pK_1 = 13.25$), but considerably higher than that of ammonia ($pK = 9.2$) or ethylenediamine ($pK_1 = 10.0$) [1, 17, 18].
3. Biguanide (I) is a stronger monoacid base, but considerably weaker than diacid bases ($pK_2 = 2.9$) such as aniline ($pK = 4.43$). This ampholytic character possibly facilitates its formation of inner metallic complexes with many transition metal ions [1,19].
4. Biguanide (I) forms stable metal complexes (six-membered ring compounds), acting as an NN-bidentate ligand, showing a strong preference for square coordination [1, 14, 20].
5. The formation constants of metal(III) biguanides are very high—even higher than those of ethylenediamine complexes [1, 19].

6. It is well established that bis(biguanide)copper(II) complexes are formed from aquacopper(II) in two steps [1], while the corresponding bis(biguanide)nickel(II) is formed in a single step [1, 21].
7. Biguanide (I) can stabilize unusually high or low oxidation states of metals, e.g., cobalt (I), manganese (IV), manganese (III), nickel (III), silver (III), silver (II), and molybdenum (III). It also forms compounds with non-metals [1, 14, 22-25].
8. Extensive magnetochemical studies of biguanide complexes confirm their inner chelate character except in the cases of those of manganese (IV) and rhenium (V) [1, 14].
9. The racemization of *levo*-tris(biguanide)cobalt(III) is the first discovered case of intramolecular twist (rhombic) racemization [1, 26-28].
10. Numerous resolutions of optically active metal biguanide complexes have been carried out, including the resolution and crystal structure of $[Co(bigH)_3]Cl_3$ [29, 30].
11. Metal complexes of biguanide (I) show interesting kinetic and mechanistic behaviour during acid hydrolysis [1, 14, 31-35].
12. Biguanide (I) does not possess any aromatic character, but when it is coordinated to any metal or nonmetal ion, the resulting chelate compound exhibits aromaticity [36,37].
13. The outer-sphere association constants of metal(III) biguanide complexes are very high [38-40].
14. The chromatographic behaviours of metal(II) and metal(III) biguanide complexes on cellulose and on silica gel are very interesting [41-47].
15. The thermodynamic aspects of the interaction of biguanide with aqueous protons have been studied and reveal that the extreme basicity of the ligand is due to the exceptional enthalpy of protonation [17, 48].
16. Biguanide acid sulphate ($C_2H_7N_5 \bullet H_2SO_4 \bullet 1.5H_2O$) is used as a primary acid-base titrimetric standard [49] and for the estimation of copper on both the macro and semimicro scales [50, 51].
17. Bis(biguanide)copper(II) chloride has been used as a reagent for the estimation of mercury [52].
18. Both biguanide and substituted biguanides are used as colorimetric reagents [53, 54].
19. Metal biguanide complexes possess antimicrobial, insect-sterilizing, and ovicidal activity [55].
20. The spectra of some octahedral metal(III) complexes have exhibited unusual aspects; normal C.D. spectra are observed in the visible region, whereas spectra in the ultraviolet region exhibit the opposite sign [56-65].
21. X-ray studies provide valuable information on the behaviour of the free ligands and their metal(II)/metal(III) complexes [29, 66-76].
22. The structure of four-coordinated metal(II) biguanide complexes is still a subject of controversy [1, 12, 76-82].

In 1961, "A Comprehensive Review of the Complex Compounds of Biguanides and Guanylureas with Metallic Elements, Inclusive of Their Many Substituted Derivatives" by Priyadaranjan Rây appeared in *Chemical Reviews* [1]. This masterly 47-page article included 188 references through December 1958. The pace of research on metal biguanide complexes has accelerated tremendously since the period covered by Rây's review, and consequently any review updating this topic must deal with a much larger number of published studies. In the present review we have tried to include all important studies published since Rây's review through 1992, including a number of later papers and communications, with particular emphasis on *synthesis, magnetism, spectroscopy, reaction kinetics,* and the *structure of the metal(II)/metal(III) biguanide complexes.* The structure of the ligands and their corresponding metal complexes will be discussed in the light of published studies of magnetic moments, electronic spectra, EPR, and X-ray crystallography.

2

GENERAL METHODS OF PREPARATION OF BIGUANIDE AND SUBSTITUTED BIGUANIDES

(i) Biguanide

Biguanide (I) may be prepared by:

(i) heating a mixture of thiocarbamide ($SC(NH_2)_2$) and guanidine ($H_2NC(=NH)NH_2$) [83];

(ii) refluxing a mixture of dicyandiamide (V) and ammoniacal copper (II) sulphate solution, whereupon rose-red bis(biguanide) copper(II) sulphate separates, which, on treatment with 10% sulphuric acid and cooling, produces biguanide sulphate crystals [84];

(iii) heating a mixture of dicyandiamide (V) and ammonium chloride [12, 85]. Ammonium methanesulphonate or ammonium benzenesulphonate may be substituted for ammonium chloride to give a satisfactory yield of biguanide(I) [86]; or best,

(iv) fusing a mixture of dicyandiamide (V) and ammonium chloride at 165°C and treating the fused mass with ammoniacal copper(II) sulphate solution, whereupon a rose-red precipitate separates, which, on treatment with dilute sulphuric acid, yields biguanide acid sulphate [87]. The free base biguanide (I) can be obtained by digesting the acid sulphate either with aqueous barium hydroxide or sodium hydroxide or by refluxing with sodium ethoxide [87, 88].

(ii) Substituted Biguanides

A large number of N^1-substituted biguanides have been prepared by fusing dicyandiamide with alkyl- or arylamine salts of methanesulphonic, benzenesulphonic, or toluenesulphonic acid. N^1-methyl, N^1-phenyl, N^1-*iso*propyl, N^1-cyclohexyl, N^1-benzyl, N^1-*p*-sulphamoylphenyl, N^1, N^1-dimethyl, N^1, N^1-diethyl, and N^1, N^1-diphenylbiguanide sulphates have been prepared by refluxing aqueous solutions of dicyandiamide, copper (II) sulphate, and the appropriate alkyl- or arylamine. The red copper compound thus obtained is dissolved in dilute sulphuric acid, and the copper is removed by precipitation with hydrogen sulphide. From the filtrate, the acid sulphate is precipitated by adding acetone [86, 89-92].

A general method for the preparation of N^1-alkylbiguanide hydrochlorides and N^1-arylbiguanide hydrochlorides is the refluxing of an alcoholic dicyandiamide solution with alkyl- or arylamine hydrochlorides [92-94] or by fusing the two reactants together in an oil bath [95]. N^1, N^1-diethylbiguanide sulphate has been prepared by fusing dicyandiamide and diethylamine hydrochloride at 130°C [93, 96].

A general method for the preparation of N^1-propyl-, N^1-*iso*amyl-, N^1, N^1-dimethyl-, N^1-hexyl-, N^1, N^1-diallyl-, N^1-crotyl-, and N^1-*iso*amylene-biguanides [97], 2-hydroxyethyl-, 3-hydroxypropyl-, 2-hydroxypropyl-, 2-methoxyethyl-, and 3-methoxypropylbiguanides [98, 99]; *iso*propyl- and *n*-butyl- [1], and benzylbiguanides [100]; hexamethylenedibiguanide [101], and piperazenedibiguanide [102] has been described by various workers. Biguanides substituted at the N^1- and N^5- positions were also prepared by the condensation of dialkylcyanamide and alkylguanidine [103].

Some N^1, N^2-disubstituted arylbiguanide bases and their hydrochlorides were prepared by the reaction of thiocarbamides and guanidine in alcoholic medium in the presence of mercuric oxide, HgO [89]. N^1, N^5-dimethyl- and N^1, N^5-diallylbiguanides were prepared by fusing a mixture of methylamine and allylamine hydrochlorides and sodium dicyanimide at 130°C. After cooling, the solid mass was extracted with water and treated with alkaline copper(II) sulphate solution, whereupon the copper compound was obtained [88]. With dimethylamine N^1, N^1, N^5, N^5-tetramethylbiguanide sulphate was obtained by the same technique [88]. N^1, N^2-dimethylbiguanide and N^1, N^1, N^2-trimethylbiguanide were prepared by the reaction of *S*-ethyl-N^1-methylguanylthiourea with methylamine or dimethylamine, respectively, in aqueous medium on a steam bath [88].

Saha and Mukherjee [104, 105] prepared needle-like white crystals of 2-guanidino-4, 6-dimethylpyrimidine (VII) (Fig. 10) and 2-phenylguanidino-4, 6-dimethylpyrimidine (VIII) (Fig. 11) by refluxing free biguanide or phenylbiguanide base with acetylacetone.

By refluxing a mixture of dicyandiamide, glycine ester hydrochloride, and cupric chloride, deep blue crystals of biguanidoethylacetato copper (II) separated. The free ligand was isolated after decomposing the complex with acid and then removing the copper with hydrogen sulphide [106]. By condensing phenylbiguanide and benzil, Saha and Bagchi prepared the Schiff base compound of composition $[C_6H_5NHC(=NH)NHC(=NH)N=C(C_6H_5)]_2$ [107].

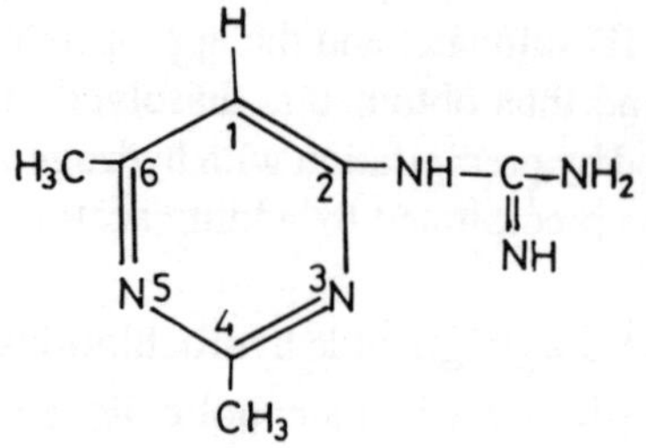

2 guanidino 4, 6-dimethylpyrimidine (VII)
Fig. 10

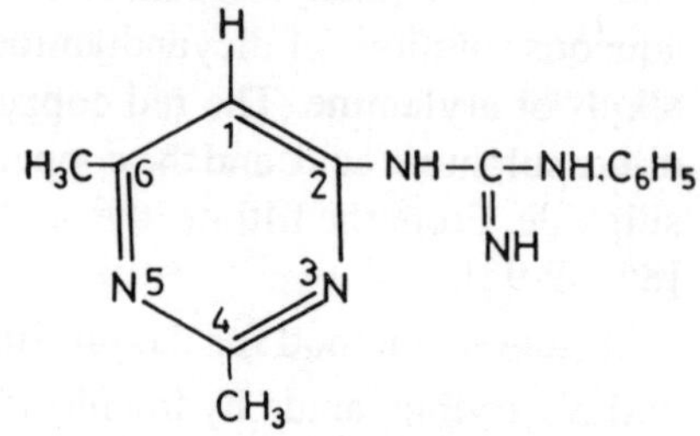

2 phenylguanidino-4, 6-dimethylpyrimidine (VIII)
Fig. 11

Recently, by the condensation of free base (methylbiguanide/ethylbiguanide) with benzil (Fig. 12), Saha *et al* [108] also prepared benzil-bis-(*N*-alkyldiamidino)diimine (IX) (Fig. 13).

$$2RNHC(=NH)NHC(=NH)NH_2 + O{=}C(C_6H_5)C(C_6H_5){=}O \longrightarrow [RNHC(=NH)NHC(=NH)N = C(C_6H_5) -]_2 + 2H_2O$$
$$(R = CH_3 \text{ or } C_2H_5)$$

Fig. 12

It is interesting to note that methyl and ethyl analogues, unlike the corresponding phenyl analogue, are neutral. This is believed to be due to the alkyl

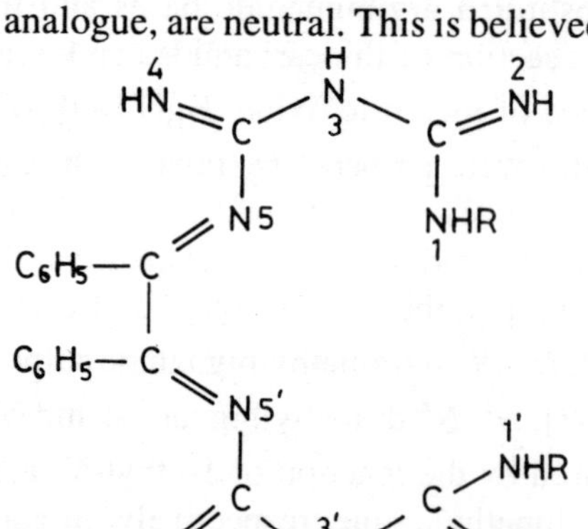

Benzil-bis(N-alkyl diamidino) diimine (IX)
Fig. 13

substitution, which hinders the possible deprotonation from N^1 and $N^{1'}$ occurring in the phenyl analogue.

The ν (C = N) band in the metal complexes shows both red and blue shifts with the free ligand (IX) (Fig. 13). Such a red shift clearly indicates that the two imine nitrogen atoms, N^5 and $N^{5'}$ coordinate to the metal. As the ligands are tetradentate, each of them further coordinates to the metal through two more nitrogen atoms, one on each biguanide part. It is the blue shift of the ν (C = N) band which, being analogous to that of the ν (C = O) band of nitrogen bound metal urea complexes, that decides those nitrogen atoms. Resonance interaction between N^1 and N^2 and between $N^{1'}$ and $N^{2'}$ lower the observed ν (C = N) band in the free ligand. If N^1 and $N^{1'}$ coordinate to the metal, a blue shift of ν (C = N) is expected, but if N^2 and $N^{2'}$ coordinate only a red shift is expected. The observed red and blue shifts of ν (C = N) thus suggest that N^5, $N^{5'}$, N^1, $N^{1'}$ coordinate to the metal.

The imine-oxime ligand

$$(C_6H_5\overset{1}{N}HC(=\overset{2}{N}H)\overset{3}{N}HC(=\overset{4}{N}H)\overset{5}{N} = C(CH_3)\check{C}(CH_3) = NOH)$$

derived from phenylbiguanide and diacetyl monoxime is reported [133]. The ν (C = N) band in copper(II) complexes shows both red and blue shifts. The red shift indicates that bonding to the metal occurs via oxime and imine $\left(\overset{5}{N}\right)$ nitrogen atoms. In the ligand, conjugation in the phenylamidine part of the type (I) (Fig. 14) lowers the ν (C = N) band to 1645 cm^{-1}. If $\overset{1}{N}$ coordinates to the metal, the ν (C = N) band is expected to show a blue shift because this mode hinders conjugation, but if $\overset{2}{N}$ coordinates, a red shift is expected. The observed blue shift of ν (C = N) clearly indicates that $\overset{1}{N}$ takes part in coordination in all complexes except nickel(II) complexes where the occurrence of only red shift of ν(C = N) indicates coordination through $\overset{2}{N}$:

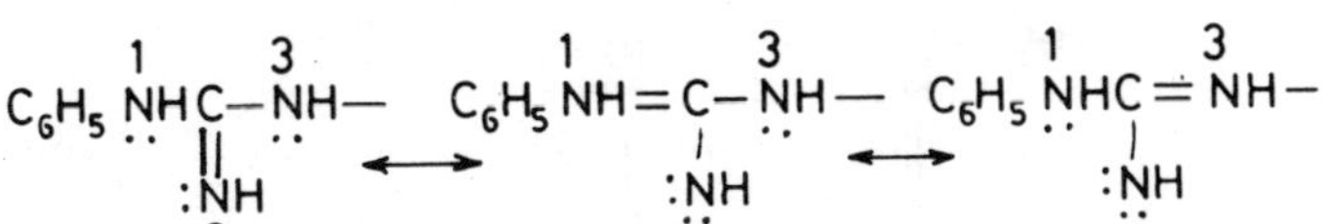

...(1)

Fig. 14

It may be noted that the frequencies of ν (N – H) bands in the metal complexes are consistently higher than that in the free ligand. This may be due to the presence of an $-NH_2$ group which might have originated from the tautomerisation of the $\overset{3}{N}H$ proton to give the NH_2 group at $\overset{4}{N}$. As the ligand in the complexes is monoprotonated, the protonation leaves the C = $\overset{2}{N}H$ group as such, which exhibits the blue shift of the ν (C = N) band. But resonance interaction of the type (2) (Fig. 15) might again result in the NH_2 group shifting the protonation to $\overset{3}{N}$.

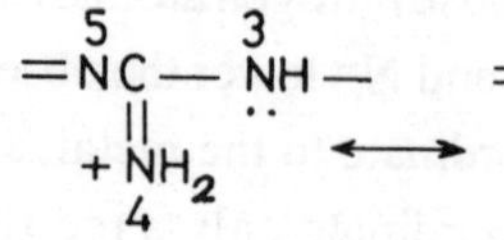 ...(2)

Fig. 15

Ethylenedibiguanide (X) (Fig. 16) was prepared by fusing a mixture of ethylenediamine hydrochloride and dicyandiamide (V). On treatment with ammoniacal copper(II) sulphate, the fused mass produces rose-red, insoluble copper(II) ethylenedibiguanide sulphate, which was isolated after decomposing the complex with dilute sulphuric acid and then removing the copper by precipitation with hydrogen sulphide [109].

Ethylenedibiguanide sulphate is now prepared in much better yield by the method determined by Chakraborty and Rây [110, 111]. Syamal and Ghanekar [112] prepared trimethylenedibiguanide sulphate by fusing trimethylenediamine dihydrochloride, dicyandiamide, and copper(II) sulphate pentahydrate at 140-145°C and finally removing copper as copper sulphide.

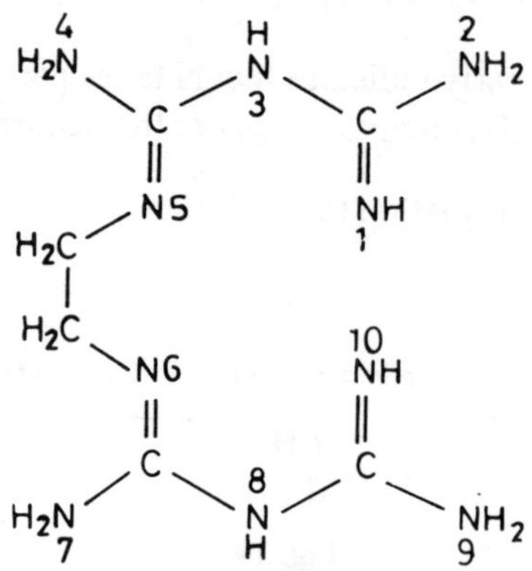

Ethylenedibiguanide (X)

Fig. 16

The ligand (X) (Fig. 16) is considered to be a true biguanide and possesses the symmetrical *cis* configuration [76, 77].

3

PROPERTIES OF BIGUANIDES

(i) General Properties, Reactions, Bond Lengths, and Bond Angles

The free base (I) can be isolated in nonaqueous solvents (methanol, ethanol, etc.). In aqueous solution, the base will be protonated; the molecule behaves structurally almost like the normal salt, as confirmed by conductance measurements [113, 114] (Fig. 17).

Salts of biguanide (I) (bigH.HCl, needle-like crystals, m.p. 235°C; bigH.2HCl, large plates, m.p. 248°C; bigH.HNO_3, prisms, m.p. 192°C;

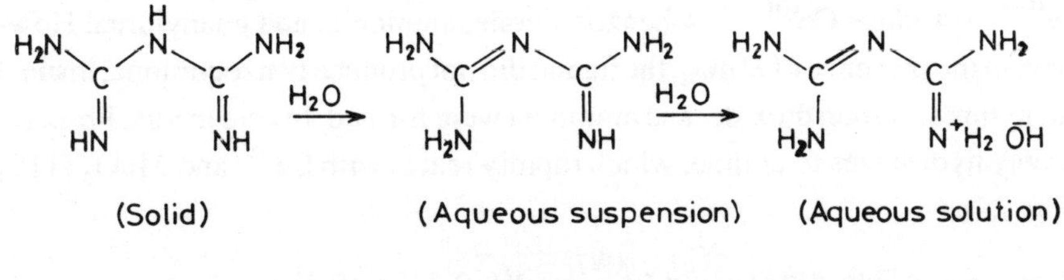

Fig. 17

bigH.2HNO_3, needles; bigH.H_2SO_4.H_2O, rhombic crystals, 231-232°C) are soluble in water and very stable in that medium [83, 87]. Some substituted biguanides also undergo facile hydrolysis in acid solution. A solution of paludrine [N^1-(*p*-chlorophenyl)-N^5-*iso*propylbiguanide] in moderately strong hydrochloric acid deposited crystals of N^1-(*p*-chlorophenyl)-N^4-*iso*propylguanylurea [115], but much more rapid hydrolysis occurs in the case of N^1-(*p*-sulphamoylphenyl)biguanide and some of its substituted derivatives [116]. In aqueous solution, biguanide obviously tends to be an equilibrium mixture of the biguanide molecule and its hydrolysis products, guanylurea and ammonia (Fig. 18).

$$\mathrm{RHNC({=}NH)NHC({=}NH)NH_2 + H_2O \rightleftharpoons RHNC({=}O)NHC({=}NH)NH_2 + NH_3}$$

$$\mathrm{p\text{-}Cl\text{-}C_6H_4\text{-}NHC({=}NH)NHC({=}NH)NHC_3H_7 \xrightarrow[H_2O]{acid} p\text{-}Cl\text{-}C_6H_4NH(CO)NHC({=}NH)NHC_3H_7 + NH_3}$$

$$\mathrm{RR'NC({=}NH)NHC({=}NH)NR''R''' \overset{+H_2O}{\rightleftharpoons} RR'N(CO)NHC({=}NH)NR''R'''\ or\ RR'NC({=}NH)NH(CO)NR''R''' + NH_3}$$

Fig. 18

The equilibrium shifts to the right in acidic media, owing to removal of ammonia as an ammonium salt.

With nitrous acid, substituted biguanide forms the corresponding dicyandiamide and guanylurea derivatives [117], and in this way, several N^4-alkyl (or N^4, N^4-dialkyl)-N^1-(*p*-chlorophenyl) guanylureas have been synthesized [115] (Fig. 19). The formation of N^1-methyl-N^1-phenylurea from N^1-methyl-N^1-phenylbiguanide and of piperidylguanylurea from piperidylbiguanide by the action of nitrous acid were also reported [118].

$$\mathrm{H_2NC({=}NH)NHC({=}NH)NHR + HNO_2 \longrightarrow H_2NC({=}NH)NH(CO)NHR + N_2 + H_2O}$$

Fig. 19

During studies of reaction kinetics, Banerjee and others [119, 120] observed that each mole of Ph-bigH/4-methoxyphenylbiguanide consumes 4 moles of $Ce^{(IV)}$ to produce $Ce^{(III)}$, 1 : 4-benzoquinone, ammonia, and guanylurea. However, in the presence of MnO_4^-, the ligand did not produce benzoquinone; instead guanylurea, carbon dioxide, and ammonia were formed. In acid media, Ph-bigH slowly hydrolyzes to aniline, which rapidly reacts with $Ce^{(IV)}$ and MnO_4^- [119] (Fig. 20).

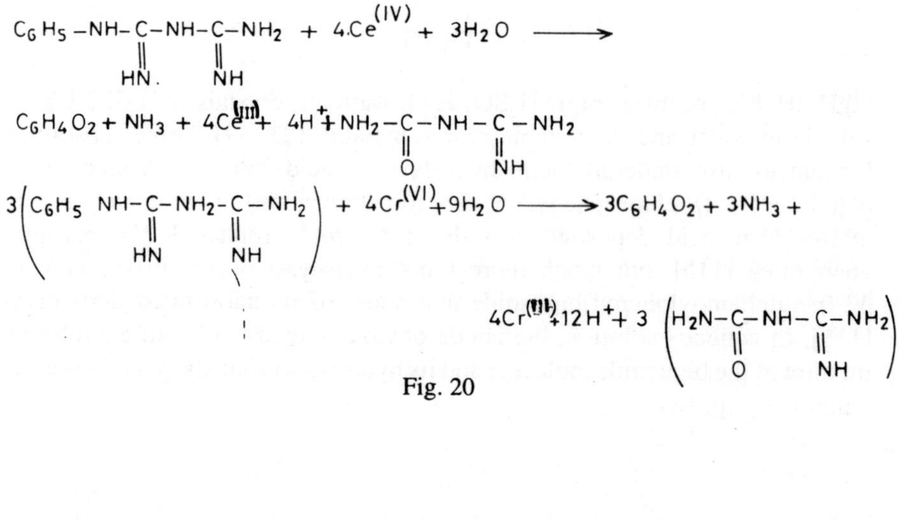

Fig. 20

Biguanide (I) is found to exist in an unsymmetrical tautomeric form [66]. The C-N bonds involving C(1) are all of intermediate length between single- and double-bond values. Unprotonated biguanide (I) (1.322 Å < C(1)-N ≤ 1.366 Å) may be compared with the C-N bond lengths reported for protonated paludrine (1.307 Å < C-N < 1.345 Å) and morpholinebiguanide (1.311 Å ≤ C-N ≤ 1.376 Å) [70, 71]. The crystal structures of free biguanide (bigH) (I) [68], bigH.HCl [67], $[bigH_2]_2[CO_3]$ [68], $[bigH_2]_2[SO_4].2H_2O$ [68], $[bigH_3][SO_4].H_2O$ [68] have been determined. The bigH.HCl molecule exists as a tautomeric form with all C-N bond lengths approximately equal.

Biguanide (I) is roughly planar, in part because of the presence of an intramolecular hydrogen bond. The twist between the two planar halves of the molecule is 12.5°. From the shortness of the C-N bonds (1.295-1.389 Å) and from the almost planar structure, it must be assumed that there is a delocalization of π-electron density throughout the molecule. The nonequivalence of the bonds may be qualitatively explained by the difference in the basicity of the various nitrogen atoms. The angles around these atoms show that the least basic ones N^1 and N^5 have some sp^3 character. Although the hydrogen bond holds the molecule in a planar configuration, it also pushes apart the two arms of the molecule in attempting to attain linearity [68].

All the ionic structures are extensively hydrogen bonded. The oxygen atoms of the anions accept multiple hydrogen bonds, one {O(2) in $[bigH_2]_2[CO_3]$} being bonded to four different hydrogen atoms, giving a distorted trigonal bipyramidal arrangement. The shortest O····H(N) distance is 1.84 Å, and the average OHN angle is 168°. In the same way that the intramolecular hydrogen bond in biguanide is a determining factor in its planar structure, the intramolecular hydrogen bonds in the ionic structures have a profound effect on the cation geometries [68].

The angle between the two planar halves of the cation are 39.5° for the chloride [67], 41.5° and 42.7° for the carbonate [60], and 44.0° and 49.1° for the sulphate [68]. This is rather surprising because the increase in the torsion angle should weaken the π-contribution to the bridge and lead to longer bond lengths. With the larger torsion angles a lengthening of one of the bridge bonds occurs, but this is barely significant. The most important effect produced by the extensive hydrogen bonding is the reduction in the thermal movement of the anions and the hydrogen atoms. This, coupled with the extensive π-bonding, gives rigid compact structures as evidenced by their higher densities and low temperature factors. The average bond lengths and bond angles of $[bigH_2]^+$ and $[bigH_3]^{2+}$ are as follows Table 1 [68]:

The C-N bonds in all the molecules have varying degrees of multiple character. In all cases π-bonding through bridging nitrogen is important but planarity in $[bigH_2]^+$ and $[bigH_3]^{2+}$ is destroyed because of steric interactions.

Table 1 Average Bond Lengths and Bond Angles for $[bigH_2]^+$, $[bigH_3]^{2+}$ and Complexed Ligands [68, 72]

Feature	$[bigH_2]^+$	$[bigH_3]^{2+}$	*[M (bigH)]*	*[M(big)]*
Bond Lengths in Å				
C–N (bridge)	1.344	1.377	1.353	1.373
C–NH_2 }	1.329	1.306	1.370 }	1.344 }
C–NHM }			1.313 }	1.305 }
Bond Angles in Degrees				
C-N-C	121.6	126.2	120.7	126.9
N-C-N (ring)	124.9	120.5	126.8	121

M(big) represents $[M\{NHC(NH_2)NC(NH_2)NH\}]$

M(bigH) represents $[M\{NHC(NH_2)NHC(NH_2)NH\}]$

Neutral bigH (Fig. 21) is planar approximately but asymmetric and intramolecularly H-bonded.

The $[bigH_2]^+$ cation (Fig. 22), determined as the carbonate and sulphate, has the same structure as that of the chloride [67]. Its structure is also quite similar

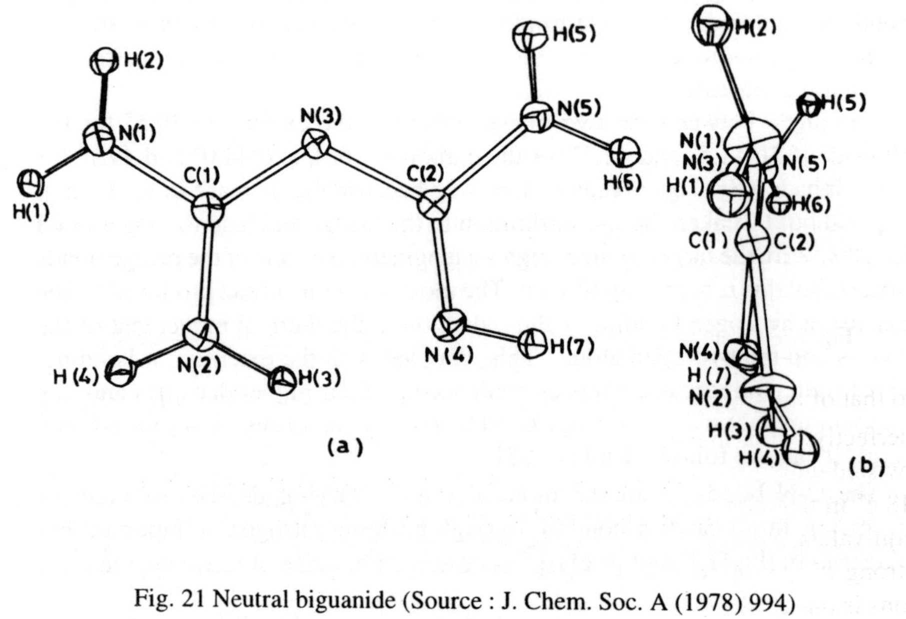

Fig. 21 Neutral biguanide (Source : J. Chem. Soc. A (1978) 994)

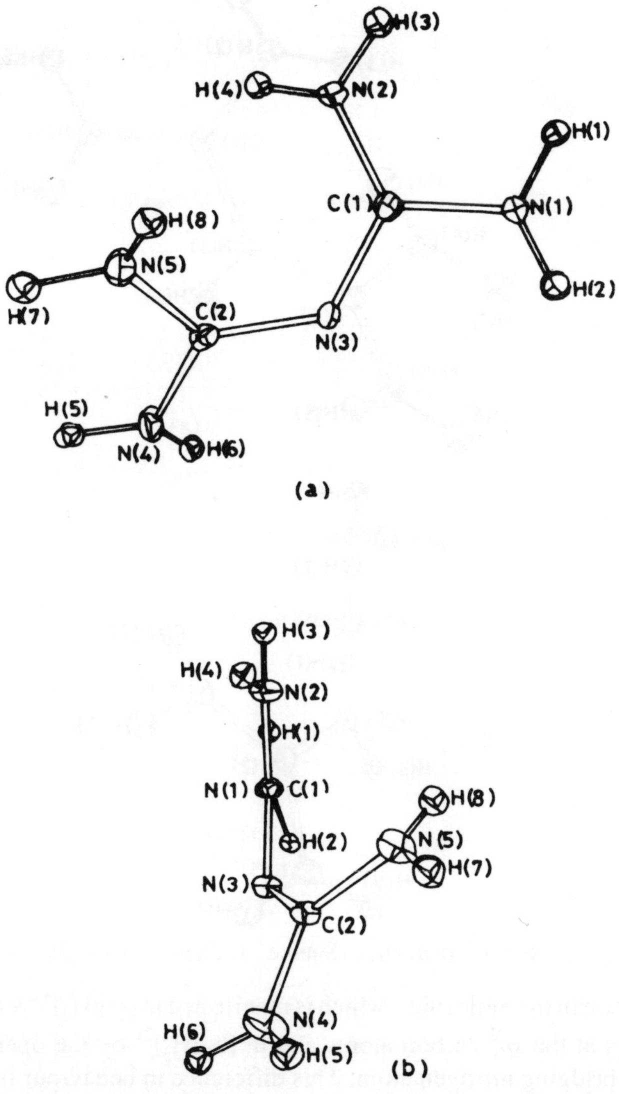

Fig. 22 Monoprotonated biguanide (Source : J. Chem. Soc. A (1978) 994)

to that of the $[bigH_3]^{2+}$ cation (Fig. 23). The ions are composed of two almost perfectly planar halves with one atom in common. The two planar parts are twisted with respect to one another by 47.6° in the monoprotonated ion and by 48.4° in the dipositive cation. The C-N bond lengths are all short and almost equivalent. From this and the planarity of the halves it is evident that there is a strong π-contribution to all the bonds. The lack of complete planarity of the ions is due to steric interaction between the hydrogen atoms. This interaction

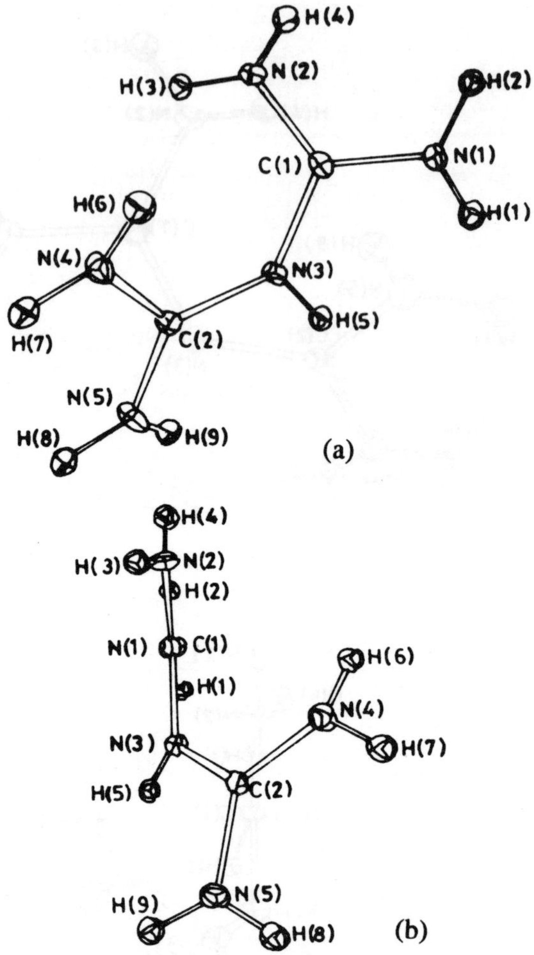

Fig. 23 Diprotonated biguanide (Source : J. Chem. Soc. A (1978) 995)

induces a strain in the molecules, which is manifested in $[bigH_2]^+$ by a distortion of the angles at the sp^2 carbon atoms and in $[bigH_3]^{2+}$ by the opening of the angle at the bridging nitrogen atom. This difference in behaviour is caused by the weakening of the bridge bonds because of the lowered basicity of the bridge atom on protonation. This weakening is also manifested by the longer C-N bridge bond (1.377 Å vs. 1.344 Å) and the shorter terminal C-N bonds (1.306 Å vs. 1.329 Å). In contrast to neutral biguanide, all the terminal nitrogen atoms are planar and sp^2 in character [68].

The crystal structures of 1-(*p*-chlorophenyl)biguanide hemihydrate, 1-(*p*-chlorophenyl)biguanide hydrochloride, 1-(*p*-chlorophenyl)-5-*iso*propylbiguanide(paludrine, Fig. 24) hydrate, 1-(*p*-chlorophenyl)-5-methylbiguanide and 1-(*p*-chloro- phenyl)-5-isopropyl-5-methylbiguanide hydrochloride hydrate, 1-(*p*-chloro- phenyl)-5-*iso*propylbiguanide hydrochloride (paludrine-hydrochloride), 1-(*p*-chlorophenyl)-5-*iso*propylbiguanide hydrobromide, and biguanide acid sulphate have been reported [121]. From the X-ray studies it is apparent that the organic cation is roughly U-shaped, and the Cl^- ion is coordinated to six nitrogen atoms at a mean distance of 3.270 Å. The biguanide part of the cation consists of two planes of atoms, each containing a CN_3 group; the planes intersect at 58.9°, and the mean C-N bond length is 1.328 Å [70]. The bond lengths and interbond angles in paludrine hydrochloride are shown in Fig. 25, XI.

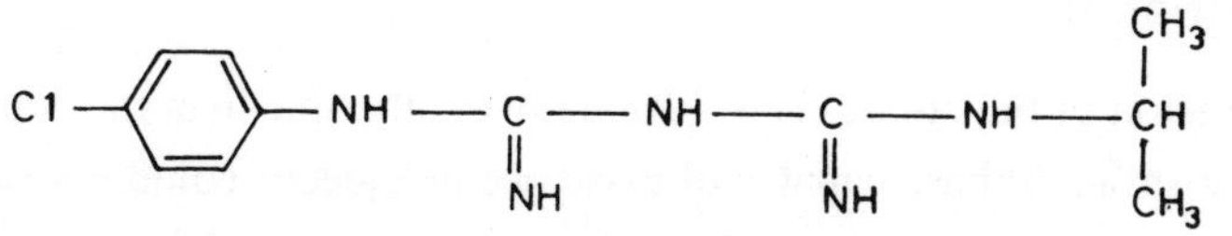

Fig. 24

During complex formation, replacement of two protons of the ligand by the metal ions occurs, removing the steric interaction that caused the deformation of the cations and allowing the ligands to approach planarity. In metal complexes, the important effect is the nonequivalence introduced into the terminal C-N bonds because of the more basic character of the MNH nitrogen atom with respect to the NH_2 group. Apart from this nonequivalence, the bond lengths and angles in analogous species are almost identical. The changes produced on protonation of the bridging nitrogen atom of $[bigH_2]^+$ also occur when the

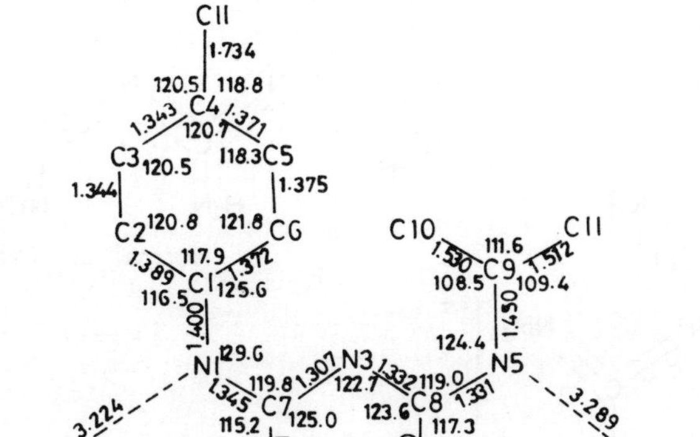

Fig. 25 Bond Lengths and Interbond Angles of Paludrine hydrochloride (XI); (Source : J. Chem. Soc. A (1967) 60)

complexed ligand is protonated. The bridge angle increases (121° vs. 127°), the bridge bonds are lengthened (1.353 Å vs. 1.373 Å), and the terminal angle at the sp^2 carbon atom decreases (127° vs. 121°). The most surprising observation is that the bridge bonds in biguanide complexes do not appear to be any stronger than in the cations even though the planarity of the complexed ligand must improve the overlap between the *p* orbital of the bridging nitrogen atom and the *p* orbitals of the adjacent carbon atoms [68].

(ii) Protonation

The protonation of biguanide (I) is 70% more exothermic than the neutralization of OH^-, the strongest base capable of existence in aqueous solution (Table 2).

Table 2 Thermodynamic Quantities of Biguanide in Aqueous Solution [17]

Process	*– ΔG, KJ/mol*	*– ΔH, KJ/mol*	*ΔS, J/mol*
$H^+ + bigH \rightleftharpoons [bigH_2]^+$	74	95	–71
$H^+ + [bigH_2]^+ \rightleftharpoons [bigH_3]^{2+}$	17	21	–13
$H^+ + OH^- \rightleftharpoons H_2O$	89	56	111
$H^+ + py \rightleftharpoons [Hpy]^+$	29	20	30

The strongly negative value of ΔS suggests that the very flexible base (I) molecule is converted into a rigid monoprotonated cation (XII) in which all the C-N bonds have pronounced double bond character:

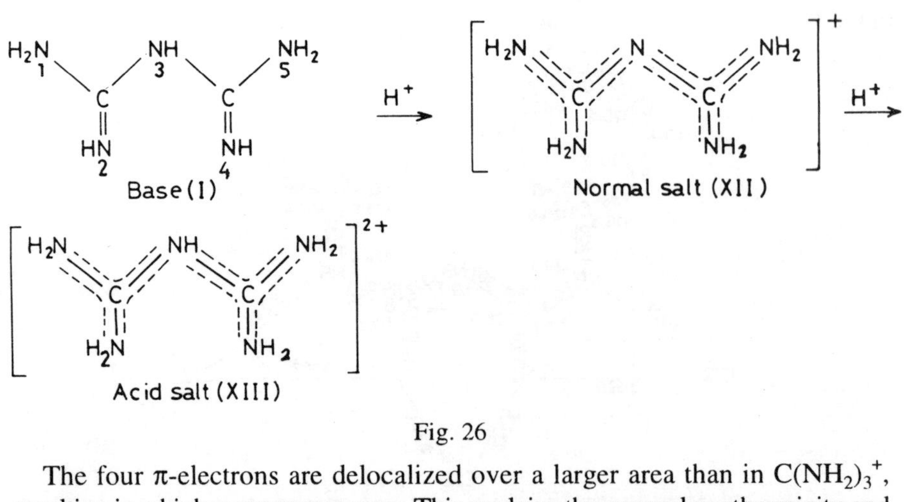

Fig. 26

The four π-electrons are delocalized over a larger area than in $C(NH_2)_3^+$, resulting in a high resonance energy. This explains the unusual exothermicity and unusually low entropy change value. The monoprotonated ligand (XII) may further accept a second proton at low pH and form a diprotonated cation $[bigH_3]^{2+}$, (XIII)) in which the second H^+ ion is linked to the bridging nitrogen atom of (XII) (Fig. 26) [17]. This is comparable to the protonation of pyridine in which the same amount of heat is produced (Table 2), which means that the singly charged $[pyH]^+$ species is less effective than the doubly charged species, $[bigH_3]^{2+}$. The free ligand containing an sp^3 nitrogen atom is puckered and thus able to arrange itself in several configurations, rotating around its σ(C-N) bonds. After complex formation the ligand becomes more rigid and loses configurational entropy.

(iii) Importance of N^3 in Biguanide

The reaction of $HgCl_2$ with an aqueous solution of $[Ni(H_2O)_2(dichlorobigH)_2]Cl_2$ leads to the formation of $[Ni(dichlorobigH.HgCl)_2]Cl_2.HgCl_2$. This reaction demonstrates that charged metal biguanide complexes are also reactive and can react with one molecule of $HgCl_2$ per bigH molecule [35]. The protons bonded to N^3 can also be replaced by silver and mercury [122]. In $[Ni(H_2O)_2\,(dichlorobigH)_2]^{2+}$, the chlorine atoms are substituted at the secondary amine nitrogen atoms (>C=NH), and hence the reaction of mercuric chloride occurs at the free $-NH_2$ group. The compound does not undergo hydrolysis or alcoholysis below 60°C. Reaction with ammonia, sodium nitrite, sodium sulphite, sulphur dioxide, or hydrazine yields the parent complex $[Ni(bigH)_2]^{2+}$.

The average bond distances and angles for the nonhydrogen atoms in $[Co(big)_3].2H_2O$ and $[Co(bigH)_3]Br_3.2H_2O$ are compared (Fig. 27). The most important point to be noted here is that upon deprotonation of the N^3 nitrogen, the C^1–N^3–C^2 angles decrease ~8°. The other slight changes upon deprotonation —the increase in the N^1–C^1–N^3 angle, the decrease in ligand bite, the decrease in the N^1–Co–N^5 angle, and the greater deviation of the ring from planarity— could all be considered to be results of decreasing the C^1–N^3–C^2 angle.

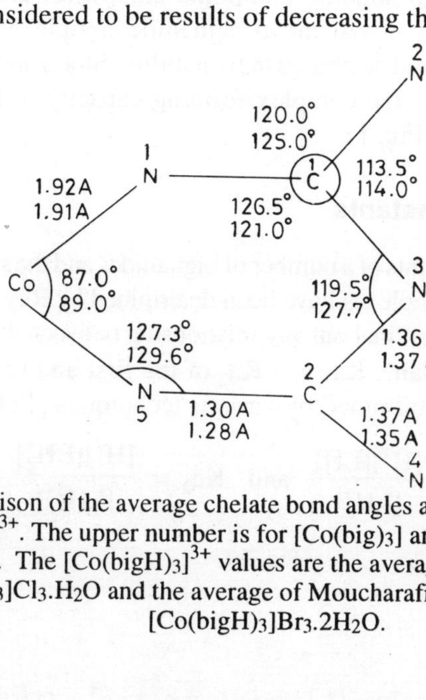

Fig. 27 Comparison of the average chelate bond angles and distances of $[Co(big)_3]$ and $[Co(bigH)]^{3+}$. The upper number is for $[Co(big)_3]$ and the lower number is for $[Co(bigH)_3]^{3+}$. The $[Co(bigH)_3]^{3+}$ values are the average of Snow's [30] values for $[Co(bigH)_3]Cl_3.H_2O$ and the average of Moucharafieh et al's [59] values for $[Co(bigH)_3]Br_3.2H_2O$.

The C-N bond distances all indicate the presence of multiple bonding, indicating the extensive delocalised π-bonding. Such strong π-bonding at the N^1–C^1 and N^5–C^2 positions are responsible for constraining the deviation of the chelate rings from planarity to the boat conformation [59].

The importance of nitrogen in position 3 (Fig. 1) for the stability of metal biguanide complexes is most spectacularly demonstrated chemically by replacing nitrogen 3 with carbon; the new compound thus formed is known as malonic diamidine (XIV) (Fig. 28).

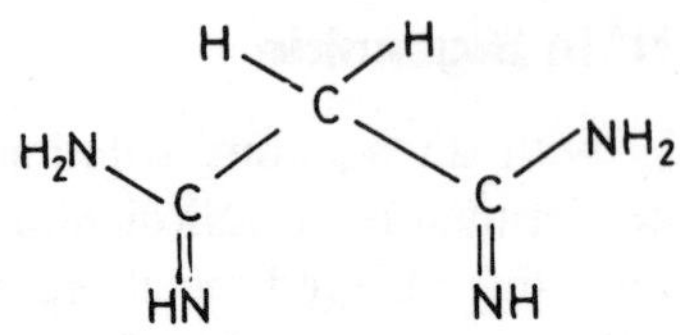

Malonic diamidine (XIV)

Fig. 28

The metal complexes of malonic diamidine (XIV) (Fig. 28) are very similar to metal biguanide complexes with respect to colours but very dissimilar with respect to stability. Bis(malonic diamidine)metal(II) [M(II) = Cu(II) or Ni(II)] complexes are transformed into metallic hydroxides in aqueous media, whereas such hydrolytic reactions do not occur with the metal biguanide complexes. In contrast to metal biguanide complexes, the malonic diamidine complexes cannot be deprotonated; decomposition of metal complexes occurs even in the presence of ethoxide base in alcoholic media. X-ray studies show that compound (XIV) is not planar; one of the carbon atoms connecting the amidine groups lies above, and the other lies below the plane of the coordination sphere [123]. Furthermore, metal-malonic diamidine complexes are not so closely packed as cationic square planar metal-biguanide complexes. This property increases their solubility and decreases their stability. Slotta and Tschesche [88] observed the inhibition of the complex-forming capacity of biguanide when substituted at position 3 (Fig. 1).

(iv) Dissociation Constants

The acid disociation constants of a number of biguanides and the stabilily constants of their many metallic complexes have been determined by Rây and co-workers [1, 124-131] with a view to final out any relationship between the two.

The dissociation constants Ka_1 and Ka_2 of the first and second protonated biguanides have been determined by various techniques [124-126] (Fig. 29),

$$Ka_1 = \frac{[H^+][LH]}{[LH_2^+]} \quad \text{and} \quad Ka_2 = \frac{[H^+][LH_2^+]}{[LH_3^{2+}]}$$

Fig. 29

where LH = one molecule of the ligand and LH_2^+ and LH_3^{2+} are the ions of the mono- and diacid salts, respectively. Ka_1 values of the first-protonated biguanide and its substituted products reveal that their basic strengths do not differ considerably from one another except in a few cases (viz., mono- and disubstituted biguanides containing an aryl, phenyl or naphthyl group). It is remarkable that bigH (pK_1 = 13.0) [17] is a much stronger base than ammonia (pK = 9.2) and ethylenediamine (pK_1 = 9.93) but is considerably weaker than diacid bases—even weaker than aniline (pK = 4.43). These properties render bigH ampholoytic, and consequently, enable it to form inner metallic complexes. Biguanides have been arranged in order of increasing basic strength (from Ka_1 values):

en(bigH)$_2$ > bigH ≈ Me-bigH ≈ Et-bigH ≈ (2-hydroxyethyl)bigH ≈ (3-hydroxypropyl)bigH ≈ (2-methoxyethyl)bigH ≈ (3-methoxypropyl)bigH ≈ Me$_2$-bigH ≈ *m*-phen(bigH)$_2$ > Ph-bigH. The dissociation constants of biguanides ($Ka_1 \approx 10^{-12}$) indicate a far stronger donor ability than guanylurea ($Ka_1 \approx 10^{-9}$) but somewhat stronger than 1-amidino-*O*-alkylureas ($Ka_1 \approx 10^{-11}$) [5, 124, 127] (Table 3).

Table 3 Dissociation Constants of Protonated Ligands

$$Ka_1 = \frac{[bigH][H^+]}{[bigH_2^+]}; \qquad Ka_2 = \frac{[bigH_2^+][H^+]}{[bigH_3^{2+}]}$$

Ligand	pK_1	pK_2	*Temperature*	*Ref.*
bigH	13.0	3.0	–	17
	11.5	2.9	32	124, 131
	12.8	3.1	25	130
Me-bigH	11.40	3.0	32	124, 131
Me$_2$-bigH	11.50	2.8	32	124
Et-bigH	11.50	3.1	32	124, 131
Et$_2$-bigH	11.70	2.5	32	124
N^1-(2-hydroxyethyl)bigH	11.50	2.8	32	126
N^1-(2-methoxyethyl)bigH	11.50	3.0	32	126
Pr^n-bigH	11.35	3.1	32	1, 127
Pr^i-bigH	11.35	3.1	32	127
N^1-(3-methoxypropyl)bigH	11.40	3.1	32	126
But^n-bigH	11.30	2.9	32	1,127
N^1-(*n*-hexyl)bigH	11.40	3.3	32	1,127
N^1-(cyclohexyl)bigH	11.40	3.2	32	1
Ph-bigH	10.70	2.2	32	124, 131
		2.25	28	131
		2.30	22	131

Contd...

Ligand	pK_1	pK_2	Temperature	Ref.
N^1-(*p*-chlorophenyl)bigH	10.40	2.2	25	130
N^1-(benzyl)bigH	11.25	2.7	32	127
N^1-(*p*-tolyl)bigH	11.80	2.6	32	127
N^1-(1-naphthyl)bigH	10.20	2.05	32	124, 127
N^1(isopropyl-N^5-phenyl)bigH	11.50	1.9	25	130
N^1-(*p*-chlorophenyl)-N^5-(*iso*propyl)bigH	{ 10.20 10.40	{ 2.3 2.3	{ 32 25	130 125
[N^1-(*p*-bromophenyl)-N^2-*iso*propyl]bigH	11.4	2.2	25	130
[N^1-(*p*-iodophenyl)-N^5-*iso*propyl]bigH	11.0	2.2	25	130
[N^1-(*p*-chlorophenyl)-N^5-*iso*propyl]bigH	12.0	2.4	25	130
[N^1-(*p*-chlorophenyl)-N^5-*iso*propyl-N^2-methyl]bigH	11.8	2.3	25	130
[N^1-(*p*-chlorophenyl)-N^5-*iso*propyl-N^4-methyl]bigH	12.2	2.0	25	5
N^1-(*p*-phenetyl)bigH	10.82	2.7	32	127
en$(bigH)_2$	{ 11.8 $(2.9)^a$	{ 11.3 1.8^b	{ 32	124
	(13.08 ± 0.2)	(12.075 ± 0.05)	25	132
m-phen$(bigH)_2$	11.4 $(2.15)^a$	9.9 $(1.4)^b$	32	124
hex$(bigH)_2$	11.5 $(3.5)^a$	10.75 $(2.3)^b$	32	127
guanylurea	8.2	1.8	32	129
AMUH	10.4	2.9	32	131

a = pK_3; b = pK_4

(v) Ultraviolet Studies

Biguanide, having a conjugated –C=N–C=N– system, absorbs at 43,000 cm^{-1} (230 nm). The addition of a single proton to this system does not shift the position of its absorption band but decreases its intensity. Such an absorption maximum at 230 nm has been attributed to a $\pi \rightarrow \pi^*$ transition. Addition of another proton causes the absorption band to disappear indicating that conjugation is no longer present in the system.

The replacement of one hydrogen atom attached to the N^1 atom by the alkyl group causes a red shift of 3-4 nm. The introduction of a second alkyl group has either no effect or may cause a shift of 1 nm (Table 4). In addition to this effect, a similar hyperchromic effect, smaller for the first alkyl group and even smaller for the second alkyl group, is expected [78, 128]. In the case of some metal biguanide complexes blue shifts have been observed [36]. In octahedral complexes an additional weak band is also observed. This is assigned to the $E \rightarrow E$ transition, split by either electronic interaction or electronic transition from an antisymmetric π-level to an antisymmetric π^*-level.

In order to ascertain the nature of the 230 nm transition Sen [78] used a simple LCAO-Hückel calculation on the skeleton (XV) (Fig. 30). The total number of orbitals is seven, and the number of π-electrons is ten (seven from the constituting atoms and three from the three NH_2 groups). The Coulomb integral parameters of N were varied. The bond orders of the various bonds are:

$$N^1\text{–}C^2 = 0.50;\ C^2\text{–}N^3 = 0.69;\ C^2\text{–}N^4 = 0.45$$

The corresponding calculated bond lengths are:

1.37 Å, 1.33 Å, and 1.38 Å

The lowest unfilled orbital is antisymmetric with respect to reflection through a plane passing through N^4. The highest orbital is symmetric, and there is one more antisymmetric and another symmetric level very close to it. The band at 230 nm may be ascribed to a transition from the symmetric to the antisymmetric level [78].

Table 4 Ultraviolet Absorption Spectra of Free Biguanides and their Salts

Compound	λ_{max} *(nm)*	λ_{max} *(cm^{-1})*	ε_{max}	*Ref.*
bigH	{230	{43,400	{15,237	134
	230	43,400	12,600	78
bigH.HCl	230	43,470	12,600	78, 128, 138
bigH.2HCl	–	–	–	78
Me-bigH.HCl	233	42,910	12,000	128
	237	42,190	12,360	135
Et-bigH.HCl	233	42,910	13,110	128
Me_2-bigH.HCl	234	42,730	12,500	128
	240	41,670	–	136
Et_2-bigH.HCl	234	42,730	12,500	128
Ph-bigH	245	40,810	–	128
Ph-bigH.HCl	245	40,810	16,010	135
	212	47,160	9,140	135
	230	43,470	11,350	135
	252	39,680	15,625	135
Cl-Ph-bigH.HCl	210	47,610	9,865	135
	232	43,100	13,760	135
	258	38,750	18,605	135
β-Ph (Et) bigH. HCl	214.5	46,620	9,052	135

Contd...

Compound	λ_{max} *(nm)*	λ_{max} *(cm^{-1})*	ε_{max}	*Ref.*
	235.5	42,460	15,890	135
paludH.HCl	211.5	47,280	13,240	135
	232	43,100	14,670	135
	258	38,750	21,350	135
	259	38,610	–	136
tolyl-bigH.HCl	213	46,940	7,592	135
	241.5	41,400	14,490	135
naph-bigH.HCl	223	44,840	41,970	135
	290	34,40	7,193	135
benzyl-bigH.HCl	233	42,910	16,390	128
morph-bigH.HCl	240	41,670	–	136
	239.5	41,750	16,370	135
guanylurea	220	45,454	9,008	134
guanylurea.HCl	220	45,454	–	134
AMUH.H_2SO_4	227	44,053	9,008	134
AEUH.H_2SO_4	227	44,053	9,387	134

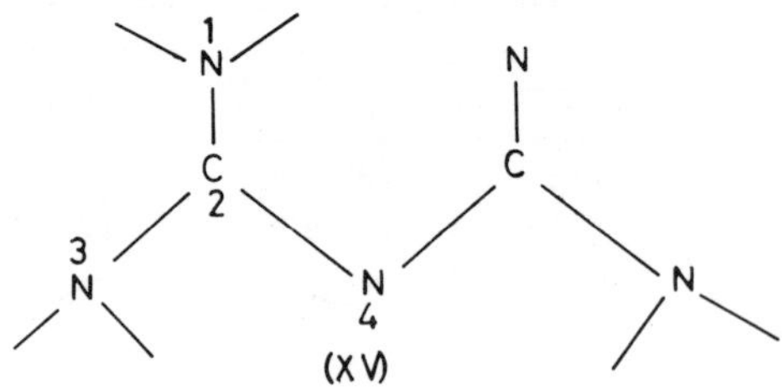

(XV)

Fig. 30

Kunchur and Mathew [76] determined the bond lengths of different C-N bonds. These values are 1.29-1.35 Å, 1.35-1.39 Å, and 1.39-1.42 Å, respectively. The N-C bonds connecting N^4 to carbon are somewhat longer than the other C-N bonds. The C-N bond length in bigH is 1.36 Å, a decrease of 0.11 Å compared to the normal C-N bond length of 1.47 Å. This contraction is partly due to the shortening of the atomic radii of the carbon and nitrogen atoms, owing to their state of hybridization, and partly to the π-bond character of the C-N bond. It can be concluded that the base (bigH) and monoprotonated biguanide $[bigH_2]^+$ molecule have a planar structure, affording the maximum opportunity for the formation of delocalized π-bonding in the entire molecule. The disappearance of the 43,000 cm^{-1} (230 nm) transition in $[bigH_3]^{2+}$ may be attributed to the pyramidal distribution of the N^4 valence bonds, destroying the planarity of the molecule. The lengthening of the bond in the Copper(II) Complex between N^4 and the two carbon atoms can be explained by the slight localization of the electron density toward the positive metal ion from the delocalized bigH molecule [78].

In metal complexes two peaks at 230 nm and 200 nm are observed. The first is due to the π-π^* transition to bigH molecule, and the second may be due to the aromatic ring containing the metal atom and the ligand. The effect of metal-ligand π-interaction shifts the $\pi - \pi^*$ transition to higher energy although some investigators had assumed that this $\pi - \pi^*$ transition should shift to lower energy since an increase in metal-ligand π-bonding involves an increase in the length of the resonance path. This rule is applicable to transitions from the highest occupied π-orbital to the lowest unoccupied π^*-orbital (which is the case for metal biguanides). Table 5 shows that the two λ max values vary from metal ion to metal ion, indicating that the overlap varies from metal to metal because of the availability of suitable orbitals.

Table 5 Ultraviolet and Visible Absorption Spectra of Some Metal Biguanide Complexes

Complex	*Medium*	*Absorption λ (nm)*	*Maxima ν (cm^{-1})*	ε	*Ref.*
$[Cu(bigH)_2]F_2$	Crystal	495 555 570	20,202 18,018 17,543		137
$[Cu(bigH)_2]Cl_2$	Crystal	495 555 580	20,202 18,018 17,241		137
	Solid		20,410		36, 78-82, 138
	Water	200 230 260 sh 520	50,000 43,478 38,400 sh 19,230	32,000 19,970 1,300 38.5	137 139 140, 141
	MeOH		18,870	43.1	75
$[Cu(big)_2].2H_2O$	Solid	500			139
	Aq. NaOH	500			139
$[Cu(bigH)_2]Br_2$	Crystal	490 545 575	20,408 18,348 17,391		137
$[Cu\ en(bigH)_2]Cl_2$	Water	520	19,230	50.5	34, 82
$[Cu(Ph\text{-}bigH)_2]Cl_2$	Water		18,200	65.5	82
$[Cu\ trime(bigH)_2]Cl_2$	Reflectance		19,720	–	112
$[Ni(bigH)_2]Cl_2.2H_2O$	Solid	445	22,472		36, 138, 139

Contd...

Complex	*Medium*	*Absorption λ (nm)*	*Maxima ν (cm^{-1})*	ε	*Ref.*
	Crystal	430 460 480	23,256 21,739 20,833		137, 139, 142
$[Ni(big)_2].2H_2O$	Solid	445			137
	Formamide	448			137
$[Ni(bigH)_2]Cl_2$	Water	204 200 232 445	49,000 50,000 43,103 22,472		31 138 139
$[Ni(Me\text{-}bigH)_2]Cl_2$	Water	465		55	31
$[Ni(Et\text{-}bigH)_2]Cl_2$	Water	465		52	31
$[Ni(Ph\text{-}bigH)_2]Cl_2$	Water	460		65	31
$[Ni\ en(bigH)_2]Cl_2$	Water	480		64	34
	Water	478		55	143
	DMSO	478		72	143
	DMF	478		62	143
$[Ni\ trime(bigH)_2]Cl_2$	Reflectance		21,500	–	112
$[Ni(paludH)Cl_2]$	Reflectance	450			
	MeOH	448 258	22,320 38,760	1.60* 4.53*	144 144
	DMF	460 260	21,740 38,460	– –	144 144
$[Ni(paludH)_2]Cl_2$	MeOH	454 258	22,030 38,760	2.08* 4.71*	144 144
	DMF	460 258	21,740 38,760	– –	144 144
$[Ni(paludH)_2(H_2O)_2]Cl_2$	MeOH	456 256	21,930 39,060	1.83 4.48	144 144
	DMF	464 260	21,550 38,460	– –	144 144
$[Ni(bigH)Cl_2]Cl$	Water	375			25
$[Pt(bigH)_2]Cl_2$	Water	220	45,500	29,250	78
$[Pd(bigH)_2]Cl_2$	Water	217 300	46,000 33,400	38,500 395	78
$[Pd\ en(bigH)_2]Cl_2$	Water	307	32,500	395	143
$[Pd\ trime(bigH)_2]Cl_2$	Reflectance		30,770	–	112

* logarithm value

Contd...

Complex	*Medium*	*Absorption λ (nm)*	*Maxima ν (cm^{-1})*	ε	*Ref.*
[Pd(Me_2-bigH) (N_3)$_2$]	Water	225 250	44,400 40,000	18,525 7,118	135 135
[Pd(morph-bigH)(N_3)$_2$]	Water	224.5 253	44,500 39,500	18,890 8,628	135 135
[Pd(Ph-bigH) (N_3)$_2$]	Water	217.5 255.5	45,900 39,100	18,690 10,953	135 135
[Pd(Cl-Ph-bigH)(N_3)$_2$]	Water	217.5 263	45,900 38,000	14,904 9,357	135 135
[Pd(paludH)(N_3)$_2$]	Water	222 261.5	45,000 38,200	24,910 14,985	135 135
[Pd(tolyl-bigH)(N_3)$_2$]	Water	219 250	45,600 40,000	16,170 7,491	135 135
[Pd(naph-bigH)(N_3)$_2$]	Water	220 250 290	45,400 40,000 34,400	15,830 4,826 3,321	135 135 135
[Pd{β-Ph(Et)-bigH} (N_3)$_2$]	Water	218 259 330	35,500 38,600 30,300	17,100 4,436 1,375	135 135 135
[Pd(morph-bigH)[$PdCl_4$]]	Water	344	29,069	–	145
[Pd(morph-bigH) [Pd(SCN)$_4$]]	Water	306 238	32,679 42,016	– –	145 145
[Co(bigH)$_3$]Cl_3	Water	190 218 266 *sh* 268 350 353 475 477	52,600 45,800 37,500 37,300 28,600 28,328 21,100 20,964	45,000 24,500 4,000 4,013 220 184 235 204	138 78 78 138 146 56 146 56
	Reflectance	360 495	27,800 *sh* 20,200	– –	147 147
[Co(big)$_3$] (anhydrous)	Reflectance	373 510	26,800 *sh* 19,600	– –	147 147
[Co(big)$_3$] (hydrated)	Reflectance	370 507	27,000 *sh* 19,700		147 147
[Co(bigH)$_3$](OH)$_3$	Water	358 482	27,932 20,746	152.1 143.2	148 148
[Co(chloro-bigH)$_3$]Cl_3	Water	386 555	25,900 18,020	665 256	36 36

Contd...

Complex	*Medium*	*Absorption λ (nm)*	*Maxima ν (cm^{-1})*	ε	*Ref.*
[Co(dichloro-bigH)$_3$]Cl$_3$	Water	386 555	25,900 18,020	665 256	148
[Co(Me-bigH)$_3$](OH)$_3$	Water	360 484	27,777 20,661	197.5 185.7	148 148
[Co(dichloro-Me-bigH)$_3$]Cl$_3$	Water	380 556	26,310 17,980	620 248	148 148
[Cr(bigH)$_3$]Cl$_3$	Water	482		103	56, 191
	Water	200 224	50,000 44,642	29,750 33,970	78, 138
	Mull	483 369		– –	
	Reflectance	483 369		– –	147
[Cr(big)$_3$] (hydrated)	Reflectance	507 385		– –	147
[Cr(big)$_3$] (anhydrous)	Reflectance	500 385			147
[Cr(big)$_3$].H$_2$O	Crystal	500 460			149
	Solid	500			139
	Water	505			139
[Cr(NH$_3$)$_6$)$^{3+}$	Water	465			150
[Cr(en)$_3$]Cl$_3$	Water	456 350		76	151
	Water	460			152
[Cr(AMUH)$_3$](NO$_3$)$_3$	Water	370 490		86 62	153
[Ru(bigH)$_3$]Cl$_3$	Water	208 260	48,076 38,500	30,500 10,550	78 78
[Rh(bigH)$_3$]Cl$_3$	Water	210	47,500	31,000	78
		307	32,500	585	78
		308	32,470		154, 155
[Ir(paludH)$_3$]Cl$_3$	Water	254 465	39,300	4.598* 3.517*	159 159
[Ir(paludH)$_2$ (NH$_3$)$_2$]Cl$_4$	Water	262 460	38,300 21,700	– –	136 136
[Ir(paludH)$_2$(py)$_2$]Cl$_4$	Water	248 460	40,300 21,700	– –	136 136

Contd...

Complex	*Medium*	*Absorption λ (nm)*	*Maxima ν (cm^{-1})*	ε	*Ref.*
$[Ir(paludH)_2]Cl_2$	Water	252	39,600	–	136
		470	21,200	–	136
$(Me_2\text{-}bigH_2)_2[IrCl_6]$	Water	236	42,300	–	136
		296	33,800	–	136
		416	24,000	–	136
		438	22,800	–	136
		484	20,600	–	136
		580	17,200	–	136
$(morph\text{-}bigH_2)_2[IrCl_6]$	Water	238	42,000	–	136
		300	33,300	–	136
		420	23,800	–	136
		436	22,900	–	136
		484	20,600	–	136
		575	17,400	–	136
$(paludH_2)_2[IrCl_6]$	Water	236	42,300	–	136
		256	39,000	–	136
		418	23,900	–	136
		436	22,900	–	136
		484	20,600	–	136
		580	17,200	–	136
$[Ir(bigH)_3]Cl_3$		322	31,250	480	158, 166
$[Ir(paludH)_2(NH_3)_2]Cl_3$	Water	252	39,600	4.513*	136, 159
$[Ir(paludH)_2(py)_2]Cl_3$	Water	254	39,300	4.515*	136, 159
$(paludH_2)_3[IrCl_6]$	Water	230	43,400	4.662*	136, 159
		256	39,000	4.792*	136, 159
		336		3.089*	136, 159
$(paludH_2)[Ir(paludH)Cl_4]$	Water	230	43,400	4.485*	136, 159
		256	39,000	4.601*	136, 159
		330	30,300	3.150*	136, 159
$(paludH_2)_2[Ir(H_2O)Cl_5]$	Water	230		4.501*	136, 159
		258		4.637*	
		340		2.920*	
$[Ir(paludH)_3]Cl_3$	Water	254		4.598*	159
		465		3.517*	159
$[Ir(paludH)Cl_2]Cl$	Water	252	39,600	4.440*	136
		470		3.321	136
$[Ag\ en(bigH)_2](NO_3)_3$	Water	448	22,321		160
		475	21,052		
		500	20,000		

* logarithm value

Contd...

Complex	*Medium*	*Absorption λ (nm)*	*Maxima ν (cm^{-1})*	ε	*Ref.*
	Polycrystaline form	450	22,222	–	160
		510	19,607	–	160
		575	17,391	–	160
		610	16,393	–	160
[Ag pipz(bigH)]$(NO_3)_3$	Mull	260			161
[VO(H_2O)(big)$_2$]	Nujol		16,700		162
			23,200		162
			25,500		162
[VO(H_2O) en(big)$_2$]	Nujol		16,700		162
			23,200		162
			25,500		162
[VO(H_2O) trime(big)$_2$]	Nujol		16,700		162
			23,200		162
			25,600		162
[VO(H_2O) tetme(big)$_2$]	Nujol		16,650		162
			23,200		162
			25,500		162
[VO(H_2O) hex(big)$_2$]	Nujol		16,700		162
			23,200		162
			25,600		162
[VO(H_2O) pipz(big)$_2$]	Nujol		16,680		162
			23,200		162
			25,500		162
[Os^{II}(Me_2-bigH)$_3$]Br_2	Water	406	24,630	3.995	163
		530	18,870	3.671	163
[Os^{II}(morph-bigH)$_3$]Br_2	Water	406	24,630	4.180	163
		530	18,870	3.462	163
[Os^{II}(paludH)$_3$]Br_2	Water	252	39,680	4.995	163
		398	25,130	3.915	163
		540	18,520	3.924	163
[Os^{III}(paludH)$_3$]Br_3	Water	256	39,060	4.520	163
		432	23,150	4.122	
		565	17,700	3.825	
[Os^{III}(paludH)$_2Cl_2$]Cl	Water	256	39,060	4.314	163
		420	23,810	3.580	
[Os^{III}(paludH)$_2Br_2$]Br	Water	254	39,370	4.275	163
		320	31,250	3.880	
		438	22,830	3.771	
[Si(H_2O)$_2$(bigH)$_2$]Cl_2	Water	231		32,180	134
[Si(H_2O)$_2$(AUH)$_2$]Cl_2	Water	219		19,380	134

Contd...

Complex	*Medium*	*Absorption λ (nm)*	*Maxima ν (cm^{-1})*	ε	*Ref.*
[B(big)$(OCH_3)_2$]	Methanol	227		13,790	165
[B(AMU)$(OCH_3)_2$]	Methanol	227		6,193	165
[Zn$(bigH)_2$]Cl_2	Water	231	43,290	13,400	166
[Cd$(bigH)_2$]Cl_2	Water	232	43,103	11,000	166
[Mn$(OH)_2$$(bigH)_2$] $(OH)_2$	Polycrystalline		19,230	–	167
			17,856	–	167
			15,923	–	167
[MnO$(bigH)_2$]$(NO_3)_4$. $2H_2O$	Water		22,200		168
			19,040		168
			17,700		168

4

METAL BIGUANIDE COMPLEXES

Biguanides (I) form highly coloured chelate compounds with a number of transition metal ions. A few reports on biguanide derivatives of main group elements such as beryllium(II), boron(III), aluminium(III), silicon(IV), zinc(II), cadmium(II), and mercury(II) have also appeared in the literature. In all cases the ligand (I) functions as a monobasic, bidentate ligand.

Rây and Kundu [169, 170] observed that 1-amidino-3-phenylurea, [$C_6H_5NHCONHC(=NH)NH_2$] which has the phenyl substituent adjacent to the $>C=O$ group, does not form metal complexes. The loss of chelating ability was attributed to the decrease in the basicity of the nitrogen-bearing aryl group. This observation is true if a methyl [88], butyl, or hexyl [171] group is in the same position. On the other hand, when the phenyl group is attached to the nitrogen atom not adjacent to the carbonyl group, [$NH_2CONHC(=NH)NHC_6H_5$] the resulting guanylurea shows complex-forming ability. Hence, the loss of chelating ability of the substituted guanylureas (substituent adjacent to the carbonyl group) is not due to the basicity of these compounds but rather to the steric factors operating between the metal atom and the nitrogen atoms with which it complexes.

(i) Structure

The structures of bis(biguanide)metal(II) [M(II) = Cu(II) or Ni(II)] complexes have been studied by many investigators to determine in which form the ligand is bonded to the metal ion. Their individual findings have created much confusion regarding the structure of metal biguanide complexes. Tschugaeff [172] proposed structure (XVI) (Fig. 31) for the bis(biguanide)metal(II) ion [M(II) = Cu(II) or Ni(II)], which was soon rejected because it cannot explain the formation and stability of the inner metallic complex.

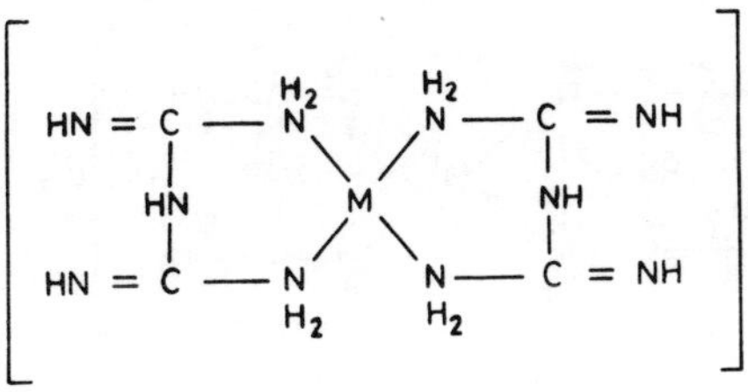

(XVI)
Fig. 31

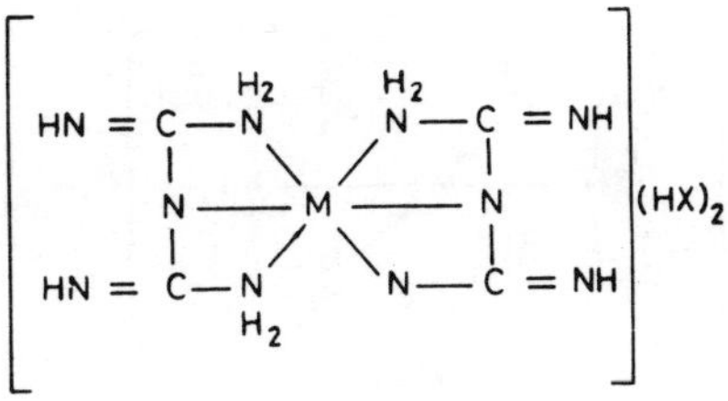

(XVII)
Fig. 32

Ley and Müller [173] formulated the inner metallic biguanide salts as structure (XVII) (Fig. 32), in which the metal atom replaces the hydrogen atom of the central imino group (=NH) of the biguanide molecule and simultaneously coordinates with the nitrogen atoms of the two amino groups. The most obvious objection to structure (XVII), is that the ligand functions as a tridentate NNN-donor and forms an unstable four-membered ring. The ligand functions as a monobasic bidentate NN-donor and forms six-membered stable complexes. The stability of these compounds in the solid state and also in solution is so great that they can be preserved for a long time without decomposition.

Slotta and Tschesche [88] noted that the complex-forming ability of biguanide is not affected by substitution in positions 1 or 5, or in both the 1 and 5 -positions (structure 1) (Fig. 1). However, if simultaneous substitution occurs in positions 1, 2, and 3 of the biguanide molecule, then a completely different character develops within the substituted biguanide in which its complex-forming ability is totally lost. The authors believed that during metal complex formation the hydrogen atom of the central imino group is replaced by the metal ion and it is this nitrogen atom, along with the two other imino groups, that is bonded to the metal ion as shown in structure (XVIII) (Fig. 33).

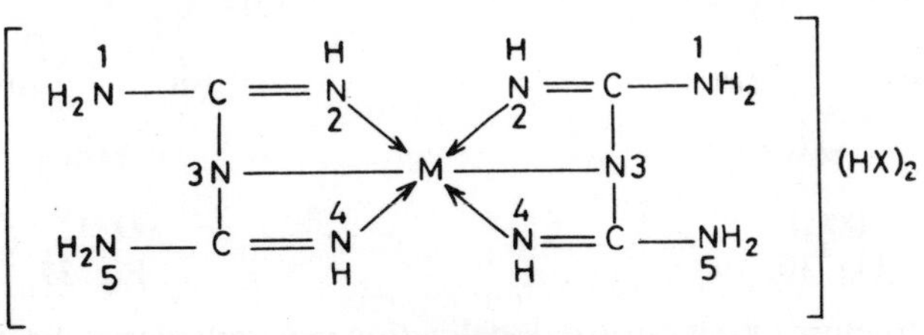

(XVIII)
Fig. 33

Structure XVIII (Fig. 33) was not accepted, as the authors assumed a four-membered ring compound. On the same grounds, the proposal of Traube and Glaubitt [174] for another four-membered ring structure (XIX) (Fig. 34) was rejected.

Rây and Saha [175] proposed that in each bigH (I) molecule the hydrogen atom of an imino (=NH) group is replaced by the metal ion and that one of the two amino groups is linked to the metal atom by a coordinate bond, thus forming

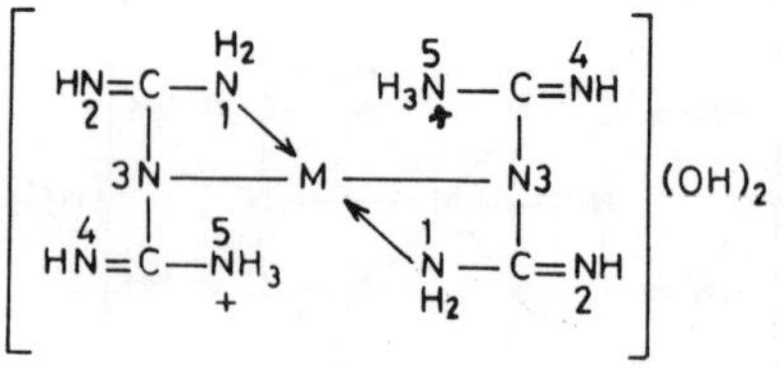

(XIX)

Fig. 34

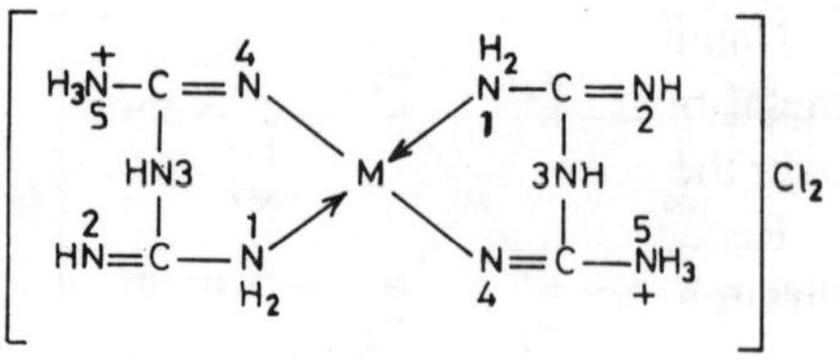

(XX)

Fig. 35

a six-membered ring (XX) (Fig. 35) [175]. The structure put forward by Rây [175] could explain many of the chemical properties, except the observation of Slotta and Tschesche [88] that the biguanide with substitution at position 3 is incapable of complex formation.

The two nitrogen atoms in position 3 (structure XX) (Fig. 35) have the largest share of the positive charge, and protons bonded to them must be the most acidic ones. Indeed, these protons of the central imino groups are lost in alkaline solution (pH ~ 12), precipitating the free, sparingly soluble, uncharged complex base. The protons bonded to the nitrogen atoms in position 3 can also be replaced by silver(I) and mercury(II) [122, 123].

The free base (I) and the normal salt (XII) are capable of complex formation, whereas the corresponding acid salt (XIII) of the same ligand requires alkali (pH ~ 12) to form metal chelates. Therefore the main requirement for complex formation has been assumed to be suitable ligand field formation around the central metal ion. On the basis of ultraviolet spectral studies, Sen [36, 78] suggested structures (XXI) (Fig. 36) and (XXII) (Fig. 37) for uncharged and cationic metal biguanide complexes respectively :

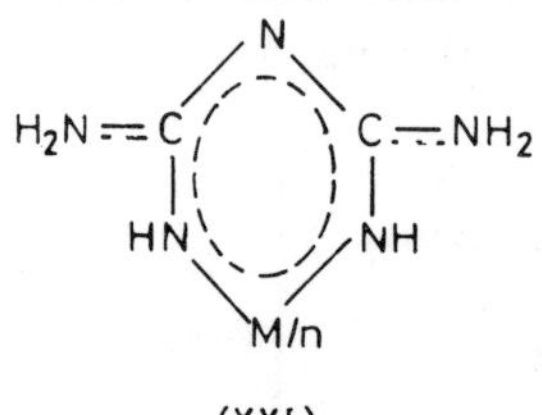

(XXI)
Fig. 36

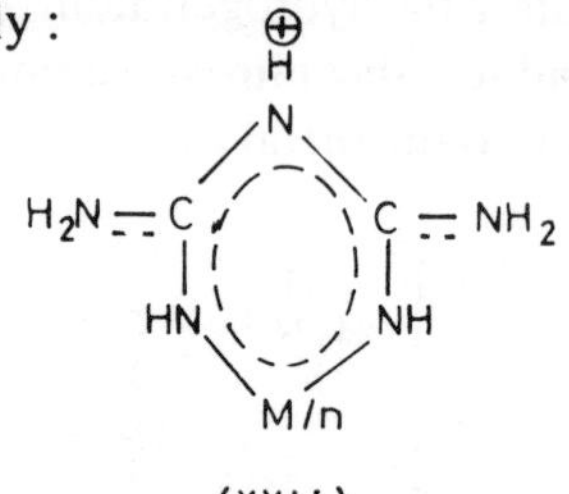

(XXII)
Fig. 37

In these structures π-electron delocalization was considered. Proton addition in the neutral complex does not disturb the planarity of the complex cation, although the proton addition to the free base ligand makes the absorption band disappear, indicating formation of the nonconjugated $[bigH_3]^{2+}$ system XIII [78]. This proton is extremely labile and can be removed by simple heating. Sen's formulae (XXI, XXII), not only readily account for the chemical aspects of metal biguanides, but they also equally explained the incapability of complex formation by N^3-substituted biguanides. His proposed structures were also supported by LCAO-Hückel calculations and N(1s) photoelectron spectral studies [176].

Nandi and Banerjee [138] also carried out ultraviolet studies of some metal-biguanide complexes and suggested that the electrons are delocalized over the entire molecule. [Fig. 36 and Fig. 37].

Recent studies have revealed that the ligand coordinates symmetrically through two =NH groups [177, 178]. In the anhydro base, (XXIII) deprotonation of the nitrogen in the 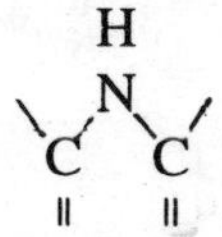 segment occurs, thus exhibiting the character of Ley's inner metallic complexes cf the first order. On careful treatment with acids, these neutral first order complexes form soluble salts (XXIV) by protonation of the negative nitrogen of the 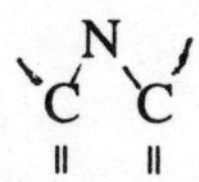 segment. Conversion of a non-electrolyte first order complex into a cationic complex due to protonation of a negative centre of the ligand yields a third order complex. Thus $[Cu(big)_2]$ (rose-red) (XXIII) (Fig. 38) is the first order inner metallic complex which changes to third order $[Cu(bigH)_2]Cl_2$ (rose-red) (XXIV) (Fig. 39) on neutralization with dilute hydrochloric acid [178].

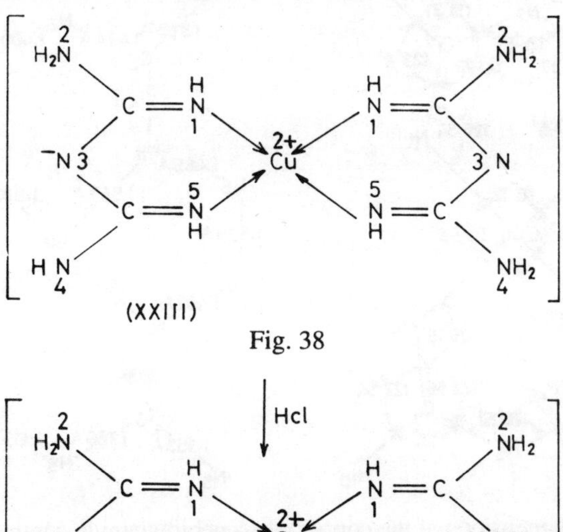

(XXIII)

Fig. 38

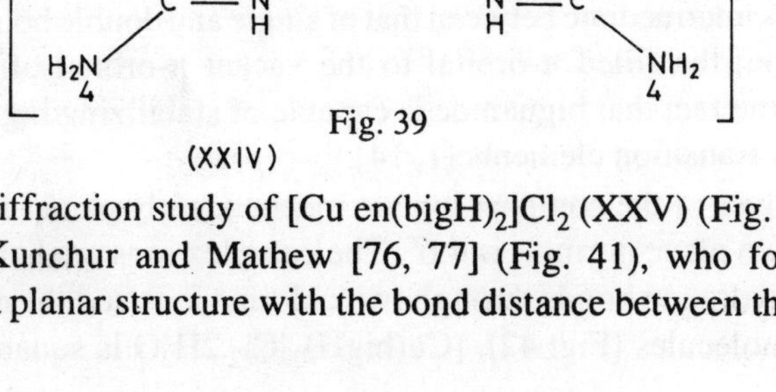

Fig. 39

(XXIV)

An X-ray diffraction study of $[Cu\ en(bigH)_2]Cl_2$ (XXV) (Fig. 40) has been reported by Kunchur and Mathew [76, 77] (Fig. 41), who found that the complex has a planar structure with the bond distance between the carbon and

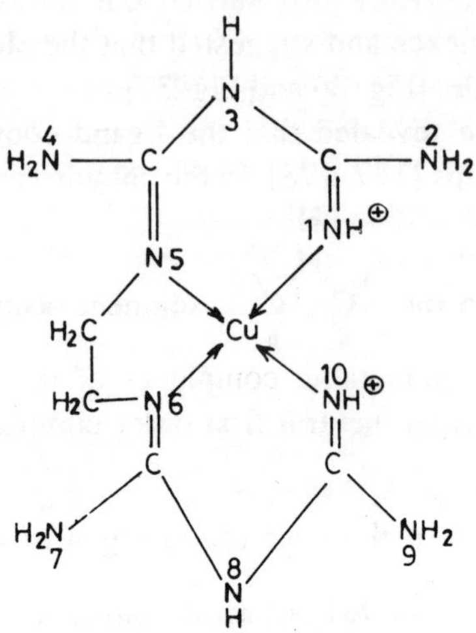

(XXV)
Fig. 40

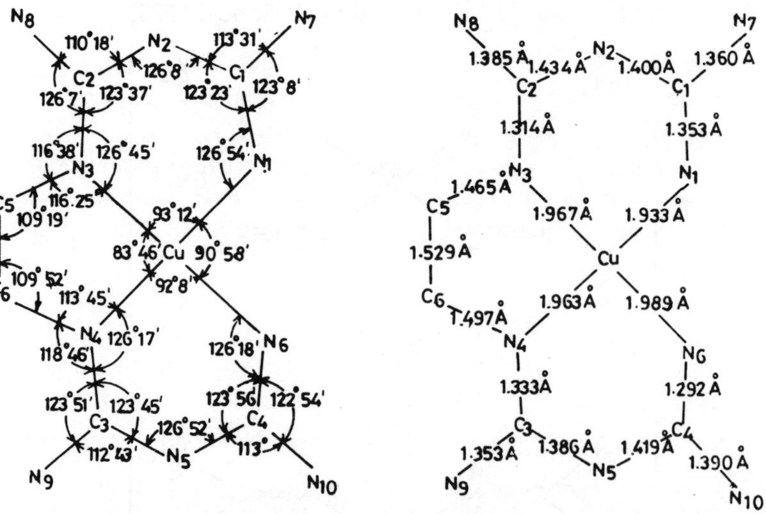

Fig. 41 Dimensions of the copper ethylenebidiguanide complex ion.
Source: *Acta Cryst.* B26 (1970) 2061

nitrogen atoms intermediate between that of single and double bounds. Electron movement from the filled π-orbital to the vacant π-orbital of the ligand is supported by the fact that biguanide is capable of stabilizing higher oxidation states of many transition elements [1, 14].

The two halves of the complex ion are independently planar, and the angle between the two plane normals is 4.6°. The complex ions are held together by a network of hydrogen bonds through two chlorine ions and the oxygen atoms of the water molecules [Fig. 42]. $[Cu(bigH)_2]Cl_2.2H_2O$ is square planar with

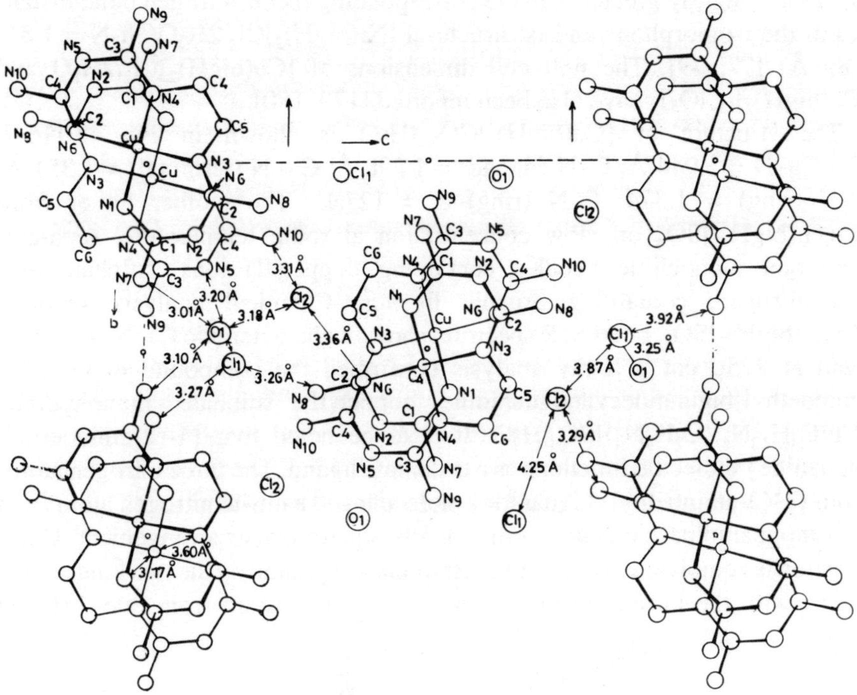

Fig. 42 A projection of the structure along the *a* axis. Distances are given in Å
Source: *Acta Cryst.* B26 (1970) 2054

the copper atom lying 0.043 Å out of the coordination plane and with a comparable Cu-N bond length (1.94 Å) [73]. The Cu-N distance and N–Cu-N angles are more regular than the values for unsymmetrical square planar [Cu en(bigH)$_2$]Cl$_2$.H$_2$O, which has a Cu-N distance of 1.933-1.989 Å. These

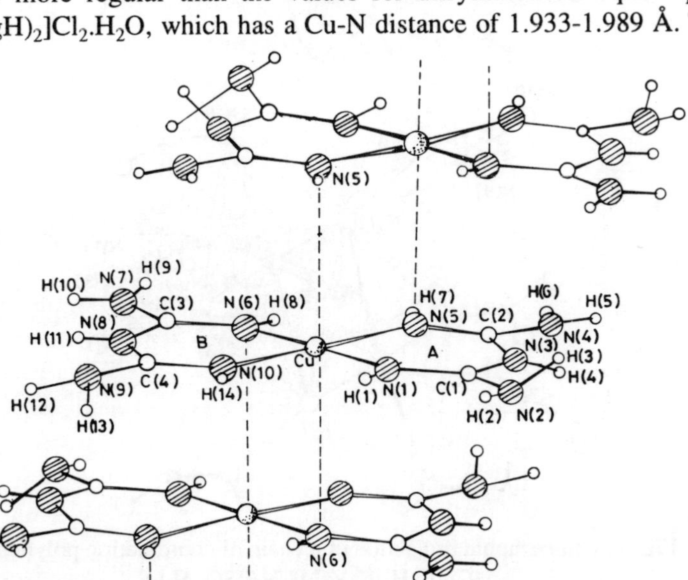

Fig. 43 [Cu(bigH)$_2$]CO$_3$ .4H$_2$O : Clinographic projection of the copper environment.
Source: *Transition Metal Chem.* 3 (1978) 69

values are slightly greater than the corresponding metal-nitrogen bond distances in the isomorphous and isostructural $[Ni(bigH)_2]Cl_2.2H_2O$(Ni-N = 1.81-1.89 Å) [77, 79]. The unit cell dimensions of $[Cu(bigH)_2]Cl_2.2H_2O$ and $[Cu(bigH)_2](ClO_4)_2$ have also been reported [179, 180].

The structure of $[Cu(bigH)_2]CO_3.4H_2O$ is shown in Fig. 43 [164]. The Cu–N = 1.944 Å; C–N (ligand) = 1.296 Å; C – N (terminal) = 1.351 Å; C–N (ring) = 1.374; C–N (ring)–C = 127.1°. The mother liquor from $[Cu\ en(bigH)_2]SO_4$ on slow concentration at room temperature, separates red-violet monoclinic needles containing copper(II) ions, sulphate ions and nitrogen, containing organic ligands. Chemical analysis supports $[Cuen(bigH)_2]SO_4$, but the IR spectrum shows a characteristic C≡N stretching band at 2250 cm^{-1}. X-ray analysis confirmed the compound to be 1-(2-aminoethyl)biguanidecyanoguanidine copper(II) sulphate monohydrate, $[Cu(C_4H_{12}N_6)(C_2H_4N_4)]SO_4.H_2O$. It is to be noted that {1-(2-aminoethyl) biguanide} molecule functions as a tridentate ligand. The three nitrogen atoms from {1-(2-aminoethyl) biguanide} molecule and a nitrite nitrogen atom from a cyanoguanidine molecule form a nearly square planar arrangement. Coordination is completed to a distorted tetragonal dipyramid by the two long copper bonds with the nitrate nitrogen and the C–N–C system from two different

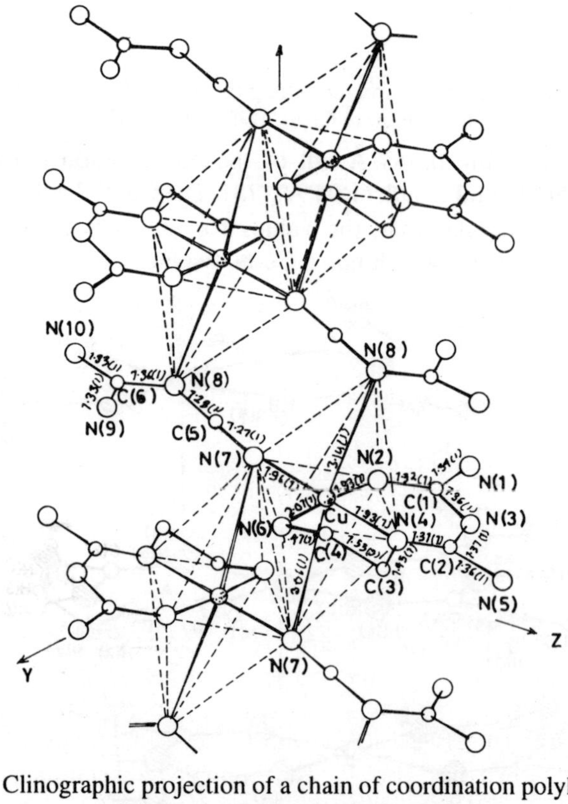

Fig. 44 Clinographic projection of a chain of coordination polyhedra in $[Cu(C_4H_{12}N_6)(C_2H_4N_9)]SO_4.H_2O$
Source: *J. Chem. Soc. Chem.* Commun., (1968) 1475

cyanoguanidine molecules which are *trans* with respect to the coordination plane [185] (Fig. 44).

By X-ray studies Wampler *et al* [79] confirmed Rây's structure (XX) [175] and ascribed minor differences to tautomeric and mesomeric effects. Interestingly, the proton transfer from one nitrogen atom to another within the ligand is extremely facile and cannot be located by ordinary chemical methods but can be traced by ^{1}H-NMR studies. Anhydrous $[Ni(bigH)_2]Cl_2$ contains 14 protons present in sets of two (position three), four (positions two and four), and eight (positions one and five) as in the free ligand [79]. The bond distances and bond angles showed that all the C–N bonds within the complex have appreciable double bond character, those within the chelate ring somewhat more so than the bonds to the noncoordinated nitrogen atoms (position 1 and 5). Thus, the charge of the complex cation resides to a large extent on the nitrogen atoms but is not located as suggested by Rây and Saha (XX) [175]. Therefore, the charged metal biguanide complex may better be represented by (XXVI) (Fig. 45).

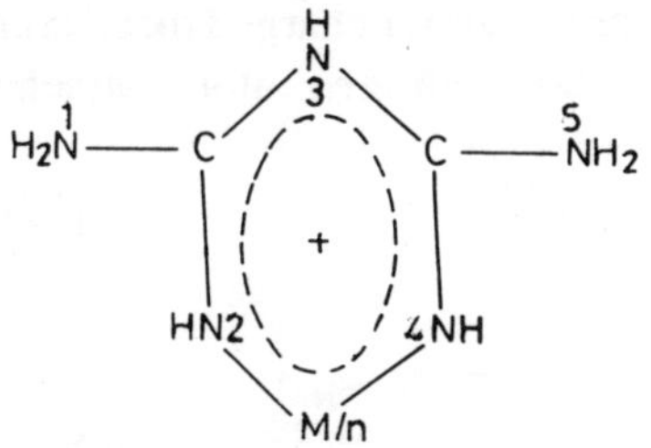

Fig. 45 (XXVI)

The N(1s) bonding energies for metal biguanide complexes (Table 6) eliminate the possible presence of the quaternary N atom in structure (XX) and support the presence of extensive π-electron delocalization over the entire molecule [176]. The NMR data for $[Ni(bigH)_2]Cl_2$ also support the absence of quaternary nitrogen atoms in charged metal biguanide complexes [79]. Therefore, structure XX proposed by Rây and Saha [175] is no longer regarded as valid.

The crystal structure of $[Ni\ en(bigH)_2]Cl_2.H_2O$ (Fig. 46) was determined by three groups—Holian and Marsh [181]; Ward, Caughlan and Smith [182]; and, Coghi, Mangia, Nardelli, and Pelizzi [184]. Ethylenedibiguanide, $en(bigH)_2$, is a tetradentate ligand and forms a square planar array about the central metal atom. The average Ni-N distance is 1.865 Å. An interesting feature is the fact that many of the hydrogen atoms of the ligand are displaced appreciably from the plane of the cation towards the hydrogen bond acceptors; the bonding about the nitrogen atoms thus becomes pyramidal. A comparison of the average bond lengths in $[Cu\ en(bigH)_2]Cl_2.H_2O$ with those in three studies of the Ni(II) analogue are shown in Table 7. The bond lengths found in isostructural $[Cu\ en(bigH)_2]Cl_2.H_2O$ are longer than those found in $[Ni\ en(bigH)_2]Cl_2.H_2O$.

Table 6 N (1s) Binding Energies for Metal Biguanide Complexes [176].

*Complex**	*Nitrogen (1s) binding energy (ev)*	*FWHM (ev)*
$[Cr(bigH)_3]Cl_3$	399.4	2.8
$[Cr(big)_3].H_2O$	399.2	3.0
$[Ag^{III}(bigH)_3]_2(SO_4)_3$	400.0	2.4
$[Cu(p\text{–}C_6H_4SO_3\text{-}big)_2]$	400.0	2.7
$[Co(bigH)_3]Cl_3$	399.2	2.7
$[Co(big)_3].H_2O$	398.9	2.9
$[Ni(p\text{-}C_6H_4SO_3\text{–}big)_2]$	400.2	2.0
$[Ni(big)_2].H_2O$	398.9	2.7
$[Cu(big)_2].H_2O$	399.5	2.9
$[Cu(bigH)_2]Cl_2$	399.5	2.5

*Nitrogen binding energies ± 0.2 eV

Single crystal X-ray studies of $[Co\ (bigH)_3]^{3+}$ and $[Cr\ (bigH)_3]^{3+}$ compounds further confirmed Sen's proposals for charged metal complexes [29, 30, 72, 78]. In $[Cr(big)_3]$ there is no hydrogen atom attached to the N^3-atom in

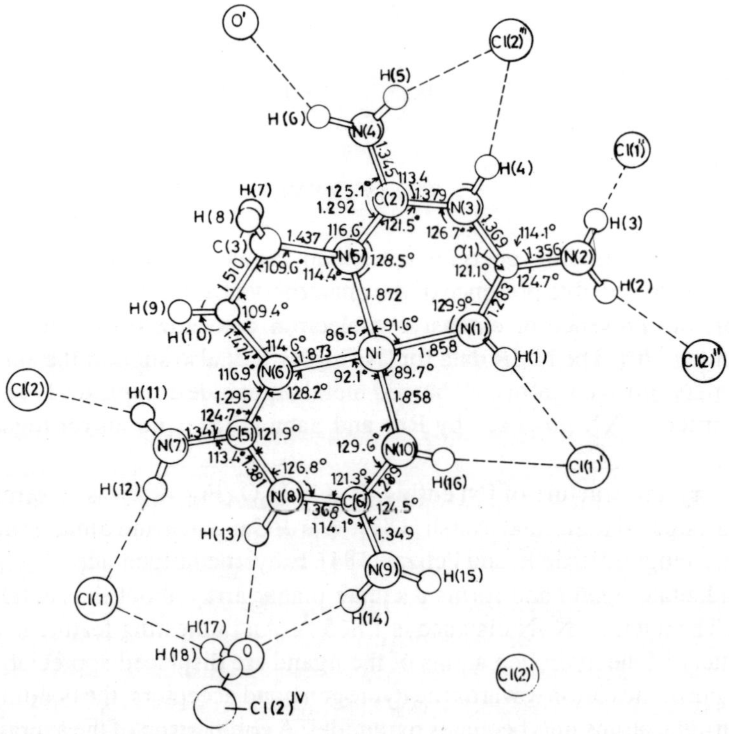

Fig. 46 Bond angles (deg) and distances (Å) within the cation. Dashed lines indicate hydrogen.
Source: *Acta Cryst.* B26 (1970) 1049

Table 7 X-ray Structural Data for Different Metal Biguanide Complexes[a]

Complex	*M–N*	*C = N (ligand)*	*C–N (ring)*	*C–N (terminal)*	*C–N (ethylene)*	*C–C (ethylene)*	*C–N–C (ring)*	*N–M–N*	*R*	*Ref.*
$[Co(bigH)_3]Cl_3.H_2O$	1.912	1.282	1.365	1.353	–	–	127.2	89.3	2.1	30
$[Co(big)_3].2H_2O$	1.926	1.301	1.356	1.374	–	–	119.0	–	–	164
$[Cr(big)_3].H_2O$	2.017	1.313	1.353	1.370	–	–	120.7	83.79	3.6	72
$[Ni(bigH)_2]Cl_2.2H_2O$	1.86	1.32	1.32	1.35	–	–	127	90.5	12.3	79
$[Ni\ en(bigH)_2]Cl_2.H_2O$	1.873	1.294	1.375	1.344	1.470	1.519	126.9	91.85	3.1	182
	1.859	1.284	1.367	1.353						182
	1.872	1.293	1.380	1.343	1.473	1.510	126.8	92.96	4.8	181
	1.858	1.286	1.369	1.353						181
	1.858	1.312	1.358	1.332	1.458	1.518	126.5	–	9.2	184
	1.853	1.295	1.359	1.355						
$[Cu(bigH)_2]Cl_2.2H_2O$	1.949	–	–	–	–	–	–	–	–	73
[Cu{1-(2 aminoethyl-bigH)} $(C_2H_4N_4)]SO_4.H_2O$	–	1.32	1.37	1.35	–	–	127.9	–	–	185
[Cu{1-(2 aminoethyl-bigH)} (NCS)]SCN	1.99	1.290	1.407	1.345	–	–	125.7	–	–	186
$[Cu(H_2O)_2(chloro\text{-}bigH)_2\ Cl_2$	2.016	1.273	1.337	1.304	–	–	124.9	82.5	–	37
$[Cu\ en(bigH)_2]Cl_2.H_2O$	1.965	1.323	1.410	1.369	1.481	1.529	126.5	92.96	–	77
	1.961	1.323	1.410	1.375	–		126.5	–	–	68
$[Ag\ en(bigH)_2](NO_4)_3$	2.00	1.31	1.40	1.30	–		125.7	88.6	–	187
$[Ag\ en(bigH)_2](ClO_4)_3$	1.97	1.32	1.36	1.35	1.50	1.54	–	–	–	23
$[Ag\ en(bigH)_2]SO_4(HSO_4).H_2O$	1.978	1.31	1.37	1.33	1.48	1.50	128.8	85.4	–	183
	1.976							92.8		

[a] The bond lengths are given in degrees, and the bond angles in Ångstrom units.
Ring C = N is given for $C–N^5$ or $C–N^1$; ring C–N for $C–N^3$; and terminal C–N for $C–N^2$ or $C–N^4$.

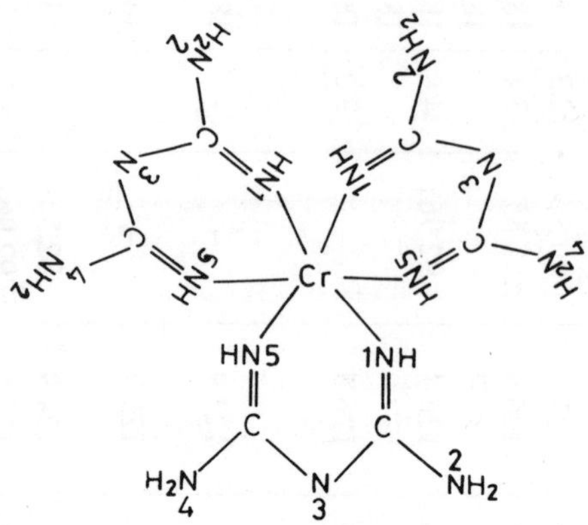

(XXVII)

Fig. 47

the $-C-\overset{3}{N}-C-$ portion of the ligand (XXVII) (Fig. 47) whereas in $[Co(bigH)_3]Cl_3.H_2O$ the protonation occurs on the N^3-atom of the $-C-\overset{3}{N}-C-$ portion of the ligand (XXVI) (Fig. 45). In $[Co(bigH)_3]Cl_3.H_2O$ the C–N bond distances follow the order: C = N (ligand) < C–N (terminal) < C–N (ring) (Table 7). This order is completely different in the case of $[Cr(big)_3].H_2O$. It may be explained that due to the deprotonation of the ring nitrogen atom in the $-C-\overset{3}{N}-C-$ part of the molecule, an increase of the π-conjugation along the $-C-\overset{3}{N}-C-$ system results, reducing the bond angle on the nitrogen atom to 120.7°, very close to the theoretical value (120°) of the angle between $-C-\overset{3}{N}-C-$. The water molecule is involved in four hydrogen bonds with nitrogen atoms of the biguanide molecules. In charged metal complexes, the $-C-\overset{3}{N}-C-$ angle lies between 124.9° and 128.8° (Table 7).

The red crystalline $(+)_{589}$-$[Co(bigH)_3]Cl_3.H_2O$ is planar with the distances Co–N = 1.912 Å; C–N (ligand) = 1.282 (2) Å; C–N (ring) = 1.365 Å; and C–N (terminal) = 1.353 Å. The data for different metal biguanide complexes (Table 7) show the following [30]:

(1) The average C–N (ligand) bond lengths range from 1.28-1.32 Å, indicating a high percentage of double bond character; these bonds remain unaltered by the deprotonation of the ligand.

(2) Unlike the C–N (ligand) bonds, the other C–N bonds are significantly influenced by the deprotonation of the molecule, which causes the C–N (ring) distances to be less than the C–N (terminal) distances. The reverse behaviour with C–N (ring) > C–N (terminal) is observed when the ligand

occurs in the neutral form, and this behaviour is also observed with 2-aminoethylbiguanide (*AEBG*) and ethylenedibiguanide en(bigH)$_2$.

(3) The C–N (ring)–C angle is the more sensitive parameter for indicating deprotonation of the ligand. For the neutral ligand (bigH) <C–N–C = ~ 127° and for the deprotonated form (big), <C–N–C = ~ 120°.

(4) In the case of the neutral ligand, the N (ligand)–C–N(terminal) angle is larger than when it is deprotonated, as a consequence of the planarity of the bonds at the carbon atom; the reverse is true for the N(ligand)–C–N (ring) angle.

(5) The N(ring)–C–N(terminal) angle is always significantly less than 120°, regardless of the form of the ligand. Yoneda *et al* [29, 74] determined the crystal structure of:
   - (i) the less soluble diastereoisomeric salt, $(-)_{589}$–[Co(bigH)$_3$]Cl (*d*-tart). 5H$_2$O
   - (ii) the more soluble diastereoisomeric salt, $(+)_{589}$–[Co(bigH)$_3$]Cl (*d*-tart). 3H$_2$O
   - (iii) the racemic chloride salt, [Co(bigH)$_3$]Cl$_3$.

In compound (i), two characteristic features were noted to be important for the mechanism of optical resolution via the less soluble diasteroisomeric salt:

(a) The chiral complex and the chloride ions are arranged alternately to form an infinite spiral chain that is cationic and left-handed as a whole.

(b) Two *d*-tart anions exist close to each other and form a dimeric unit with four negative charges, structurally similar to the antimony *d*-tartrate ion. Such a dimeric unit of the *d*-tartrate ion has not yet been found in crystal structure determinations.

A similar infinite spiral chain was found in the optically active chloride salt $(+)_{589}$–[Co(bigH)$_3$]Cl$_3$.H$_2$O. In this case, the spiral chain consists of Δ complexes and is right-handed as a whole.

In compound (ii), the complex cation and the chloride anions are arranged alternately to form an infinite spiral chain, $\{\Delta$-[Co(bigH)$_3$]Cl$\}_\alpha$. Each *d*-tartrate ion is hydrogen bonded to a two-fold-screw-axis related neighbour (–OH .... OOC = 2.792 Å) to form an infinite chain, (*d*-tart)$_\alpha$.

In compound (iii), two kinds of similar infinite chains $\{\Delta -$ [Co(bigH)$_3$]Cl$\}_\alpha$ and $\{\Lambda -$ [Co(bigH)$_3$]Cl$\}_\alpha$, which are mirror images of each other, are found. The structures of (ii) and (iii) are compared with those of the less soluble diastereoisomeric chloride *d*-tartrate and the optically active chloride, in both of which a similar infinite chain has been found.

The crystal and molecular structure of [Cr(big)$_3$].H$_2$O was determined by Coghi, Nardelli, and Pelizzi [72] (Fig. 48). The main feature of this compound is the deprotonation of the biguanide molecule, which produces an enhanced π-conjugation along the $-\overset{1}{C}-\overset{3}{N}-\overset{2}{C}-$ system, reducing the bond angle on the nitrogen atom to the theoretical value of 120°. The chromium atom is surrounded by three bidentate biguanide molecules, giving rise to an octahedral arrangement. The Cr-N bond lengths are all equivalent (1.98-2.16 Å), and the N–Cr–N bond angles, which involve N atoms belonging to the same biguanide ligands, are all nearly equal and significantly smaller (83.04°, 84.22°, 84.11°) than the others (88.54° to 95.68°). This is a consequence of the rigidity of the ligand, which shows the same "bite" distance (N .... N ≈ 2.7 Å) and coordinates to the metal through equal N–Cr interactions.

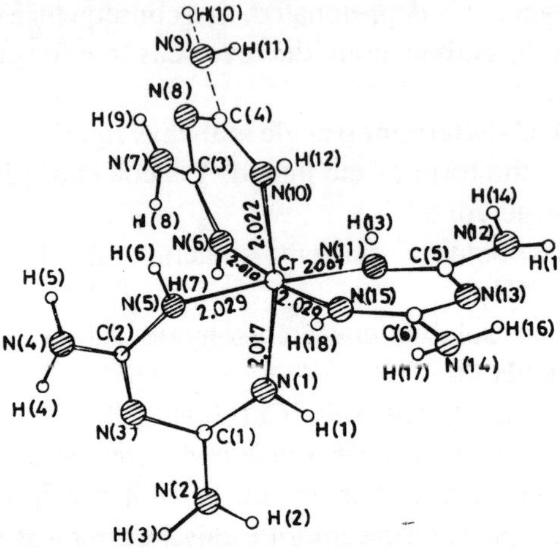

Fig. 48 Clinographic projection of the structure.
Source: *Acta Cryst.* B32 (1976) 842

The results of Coghi *et al* show that for the protonated ligand the following sequence is observed:

C–N (ligand) < C–N (terminal) < C–N (ring)

In the case of the deprotonated ligand this order is inverted for the last two bonds, which produces an increased π-conjugation along the central $\overset{1}{-C}-\overset{3}{N}-\overset{2}{C}-$ system in the ligand, thus reducing the bond angle on the nitrogen atom to the theoretical value of 120°. (In the case of the protonated ligand this angle ranges from 125.7° to 128.8°). In [Cr(big)$_3$].H_2O, the water molecule is involved in four hydrogen bonds with the nitrogen atoms of the ligand molecules (Fig. 48). Brubaker and Webb [57] determined the crystal structure of (–)$_D$–[Cr(bigH)$_3$] by X-ray diffraction of the least soluble *d*-10-camphorsulphonic acid trihydrate salt. Unfortunately, the authors did not report the molecular geometry of the cation. The cell dimensions of metal(II)/metal(III) biguanide complexes have also been reported [23, 29, 37, 57, 59, 72, 75, 77, 79, 179-183, 186-190]. In the [Cu en(bigH)$_2$]$^{2+}$ cation (XXV) the non-planarity of the two biguanide residues can be explained as resulting from the strain in the five-membered ring containing the two ethylene groups [77].

In [Ni en(bigH)$_2$]Cl_2 the ligand wraps itself around the Ni atom so as to form a square planar array of Ni-N bonds (Ni-N = 1.865 Å). Despite considerable double bond character in the C–N bonds, many of the H atoms are displaced

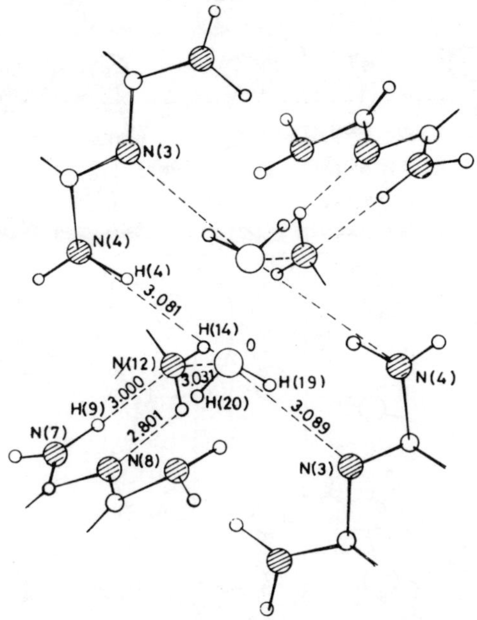

Fig. 49 Hydrogen bonds involving the water molecule.
Source: *Acta Cryst. B32 (1976) 842*

appreciably from the plane of the cation towards the H-bond acceptor; the bonding about the N atoms thus becomes pyramidal. On heating the compound in air at 150°C, the crystal begins to decrepitate and become opaque, presumably as the water of crystallization is driven off. At 260°C the crystals begin to darken, and at about 330°C they become dark. At a higher temperature (350°C) the sample becomes black but remains solid, indicating that the crystals are held together by an extensive network of H-bonding. All the available hydrogen atoms form H-bonds either to Cl^- ions or to water molecules [181].

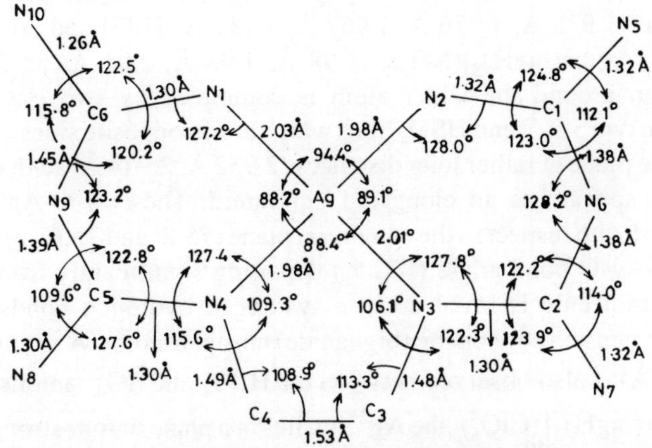

Fig. 50 Bond lengths and bond angles of ethylenebidiguanide silver(III) nitrate.
Source: Nature 217 (1968) 540

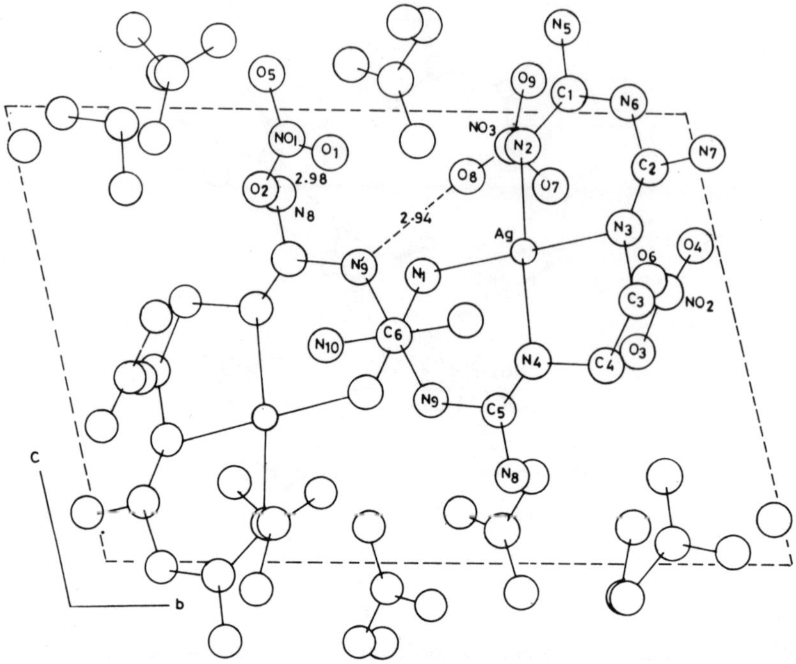

Fig. 51 A diagrammatic representation of the structure projected down the *a* axis
Source : *Nature* 217 (1968) 540

Although the average metal-nitrogen bond lengths of [M en(bigH)$_2$]Cl$_2$ [M = Cu(II) or Ni(II)] and [Ag en(bigH)$_2$]X$_3$ (X = NO$_3$, ClO$_4$ or 1/2SO$_4$) are essentially the same, in the silver(III) complex the C–N bond length indicates more uniform multiple bonding than in the case of the nickel(II) and copper(II) analogues, indicating that the highly charged silver(III) ion is accommodated by extensive delocalization over the entire biguanide structure. Figs. 50 and 51 show the bond lengths and bond angles of [Ag en(bigH)$_2$]$_3$(NO$_3$)$_3$ [187].

The four Ag-N bond distances of [Ag en(bigH)$_2$](HSO$_4$)SO$_4$.H$_2$O (Fig. 52) are all equal (1.978 Å, 1.976 Å, 1.967 Å, 1.987 Å) [183] and are very close to those in [Ag en(bigH)$_2$](NO$_3$)$_3$ (1.98 Å, 1.98 Å, 2.01 Å, 2.03 Å) [187]. Coordination around the silver atom is completed by two oxygen atoms belonging to two SO_4^{2-} and HSO_4^- ions which are on opposite sides with respect to the square plane at rather long distances (2.932 Å, 2.944 Å) so that the entire polyhedron appears as an elongated bipyramid. The two O-Ag bonds are equally tilted with respect to the equatorial plane (75.2° and 75.6°, respectively) with the O-Ag-O bond angle (152.5°) departing significantly from linearity. The water molecule is involved in a system of hydrogen bonds involving sulphate oxygen and ethylenebis(biguanide) nitrogen atoms. A strong hydrogen bond (2.51 Å) is also observed between the HSO_4^- and SO_4^{2-} anions (Fig. 53).

In [Ag en (bigH)$_2$] (ClO$_4$)$_3$ the Ag^{3+} ion lies in a plane of four strongly bonded N atoms with the Ag^{III} –N bond distance = 1.97 – 1.99 Å. This is ~ 0.18 Å less

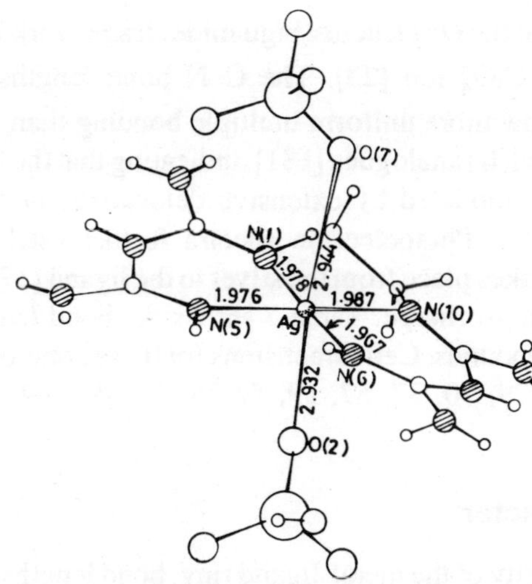

Fig. 52 Clinographic projection of the structure with the atomic distances in the coordination polyhedron.
Source: *Acta Cryst.* B31 (1975) 131

than the Ag^{II} –N bond distance, and the Ag^{3+} ion is weakly bonded to a perchlorate oxygen at 2.82 Å above the plane. The closest atom from a neighbouring $[Ag\ en(bigH)_2]^{3+}$ fragment is N^8 at 3.54 Å. Each biguanide unit

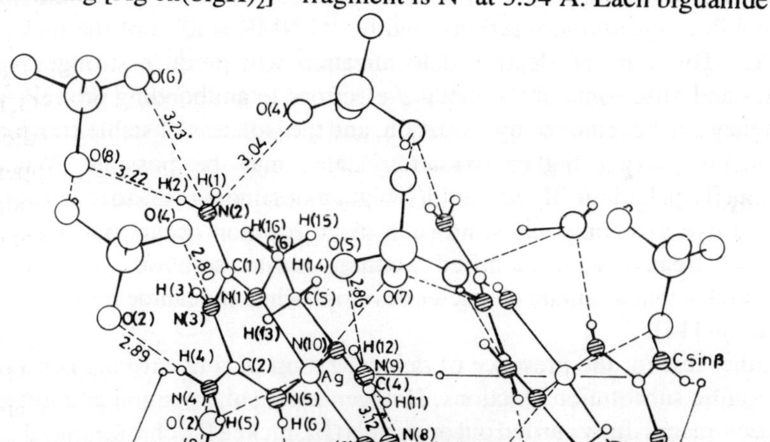

Fig. 53 Diagrammatic projection of the structure of ethylene bis(biguanide) silver(III) sulfate hydrogen sulfate monohydrate, perpendicular to [100], showing crystal packing atom bonding. Source: *Acta Cryst.* B31 (1975) 131

is essentially planar, but the ethylenebis(biguanide) framework is bowed away from the coordinated ClO_4^- ion [23]. The C–N bond lengths found in the silver(III) complex show more uniform multiple bonding than in the cases of the copper(II) and nickel(II) analogues [181], indicating that the highly charged silver(III) ion is accommodated by extensive delocalization over the entire biguanide structure [23]. Photoelectron spectra further establish that little charge delocalization takes place from the silver to the ligand [176]. This leaves silver with a high positive charge. Table 7 shows the bond lengths and bond angles of different compounds. Cell dimensions for these compounds have also been determined [23, 29, 30, 37, 57, 59, 72-75, 77, 79, 164, 172-182, 189, 187-191].

(ii) Aromatic Character

Facts such as the planarity of the metal-ligand ring, bond lengths lying between those of single and double bonds, and metal-ligand ring systems containing 6 π-electrons point to the possibility of aromaticity in the ring. Biguanide contains filled π-orbitals which have the same symmetry as some *p-d* orbitals of the metal ions, and these can result in complete delocalization of electrons in the metal-biguanide chelate ring. Such π-electron delocalization in metal chelates has been supported by LCAO–MO calculations, by a number of electrophilic substitution reactions, and by ^1H NMR studies of the nickel(II) complex. This type of electron delocalization will produce stronger metal chelates and raise some of the metal *d*-electrons to antibonding orbitals from which they can be removed by oxidation, and the isolation of stable complexes with metals having higher oxidation states may be possible. Whereas platinum(II), palladium(II), and gold(I) biguanides undergo diazo reactions and form red dye azo compounds, no such diazo reaction occurs when the free ligand is treated with acidified sodium nitrite solution. This clearly demonstrates that aromatic character develops when biguanide coordinates to a metal ion [14].

Another test for the presence of delocalization of ring currents is positive electrophilic substitution reactions. Halogenation, sulphonation and nitration have been successfully carried out on copper(II), nickel(II), chromium(III), and cobalt(III) complexes. Preparation of robust, water-stable boron and silicon compounds with these ligands indicates electron delocalization in the ring system. The band (~ 200 nm) assigned to aromatic metal chelating systems is red-shifted in their halogen derivatives [36, 78]. A decrease of 10 Dq and an increase of β values are noted in halogen-substituted complexes compared to simple complexes. The presence of a halogen atom in the ligand weakens the σ- and π-donor capacities of the nitrogen atoms and increases the length of the M–N bonds, making the compounds less stable [36]. For substituted and unsubstituted copper(II) complexes, a lengthening of the metal-nitrogen bond nearer the halogen atom and a shortening of the $\equiv$ C–NH_2 bond have been

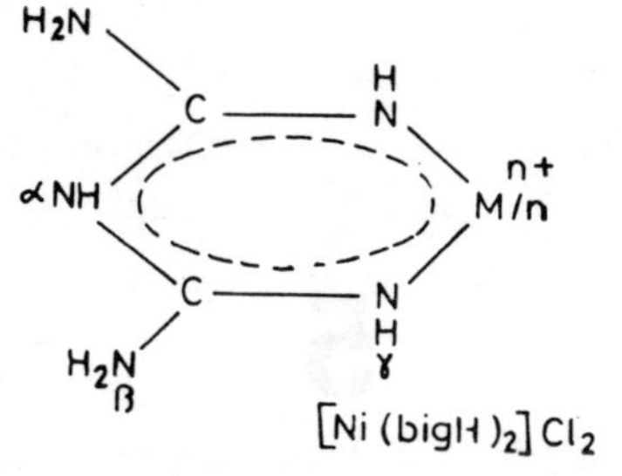

(XXVIII)

Fig. 54

attributed to a drifting of the delocalized electron cloud towards the halogen atom.

Creitz *et al* [79] measured the C–N bond distance (1.32 – 1.35 Å) and bond angles (XXVIII) (Fig. 54) of $[Ni(bigH)_2]Cl_2.2H_2O$ (Table 7), which is consistent with Sen's π-electron delocalization [36, 78]. ^{1}H NMR studies of anhydrous bis(biguanide)nickel(II) chloride (XXIX) (Fig. 55) indicate τ values of 0.12 for N(α), 3.4 for N(β), and 5.35 for N(γ). The donor field chemical shift

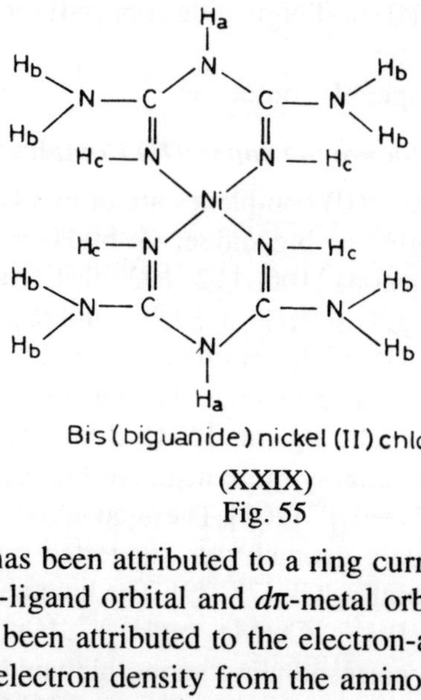

Bis (biguanide) nickel (II) chloride

(XXIX)

Fig. 55

of 0.12 for N(α) has been attributed to a ring current effect arising from the π-cloud by the *p*π-ligand orbital and *d*π-metal orbital. The chemical shift of 3.40 for N(β) has been attributed to the electron-attracting >C = NH group, which withdraws electron density from the amino nitrogen atom and in turn from the protons. The high field positions of the N (γ) with respect to the other two peaks may be attributed to shielding by the d_{xy} orbital of the diamagnetic d^8-metal ion.

From X-ray and NMR reports the structure of bis(biguanide)nickel(II) chloride may be depicted as in (XXIX) (Fig. 55).

5

BIGUANIDE CHELATES

(I) Copper(II) Complexes

These compounds are subdivided under the following headings:

(a) bis(biguanide)copper(II) or dibiguanide copper(II) complexes
(b) copper(II) heterochelates
(c) mono (biguanide) copper(II) complexes.

(a) Bis(biguanide)- and Dibiguanide copper(II) Complexes

The rose-red bis(ligand) copper(II) complexes are of two types: $[Cu(R\text{-}big)_2]$ and $[Cu(R\text{-}bigH)_2]X_2$ (bigH = biguanide; R-bigH = alkylbiguanides) (X = OH or any univalent anion) [100, 112, 192-203]. The anhydro base is formed by heating $[Cu(R\text{-}bigH)_2](OH)_2$ at 100°C, and the salts are prepared by heating $[Cu(R\text{-}bigH)_2](OH)_2$ with ammonium salts [1, 5]. The formation of $[Cu(bigH)_2]^{2+}$ is more exothermic than the reaction of copper(II) with strongly coordinated macrocyclic ligands [204]. Similarly, the formation of $[Ni(bigH)_2]^{2+}$ involves a remarkably more negative enthalpy change than the reference amine complex $[Ni(en)_2]^{2+}$ [205]. The aqueous complexes presumably have a tetragonal or square geometry as suggested by X-ray studies of $[Ni(bigH)_2]Cl_2$ and $[Cu\ en(bigH)_2]Cl_2$ [77, 79]. The kinetics of dissociation of $[Cu(bigH)_2]^{2+}$, $[Cu(Me\text{-}bigH)_2]^{2+}$, $[Cu(Me_2\text{-}bigH)_2]^{2+}$, $[Cu(Et_2\text{-}bigH)_2]^{2+}$, and $[Cu(Ph\text{-}bigH)_2]^{2+}$ into the corresponding mono(biguanide)copper(II) complexes ($pH = 4.5 - 5.5$) have been studied by Banerjea [206, 207].

All bis(bigH)/(substituted bigH)copper(II) complexes have magnetic moments 1.73-1.84 B.M. and show a spectral band at approximately 19,000 cm^{-1} in water and 20,000 cm^{-1} in the solid state (Table 5).

Electronic, ir spectral and thermal studies of some copper(II) complexes were reported by Babykutty and others [139, 208]. The water of hydration in $[Cu(bigH)_2]Cl_2.2H_2O$ was eliminated at 100-125°C, followed by and endothermic decomposition at 190-210°C. Above 400°C, the compound totally decomposed to CuO. In $[Cu(bigH)_2]SO_4.3H_2O$ the water of crystallization begins to be lost at 40°C, and anhydrous $[Cu(bigH)_2]SO_4$ is stable between 94-146°C. A mechanism supporting the intermediate formation of CuS and the ultimate formation of CuO above 830°C has been suggested [208].

Compounds of types $[Cu(R\text{-}bigH)_2][CuR'_2]$ and $[Cu(R\text{-}bigH)_2][R''H]_2$ (R-bigH = bigH or Ph-bigH) (R'H = *N*-benzoylsulphonylglycine, *N*-benzoylsulphonyl-*dl*-α-alanine, *n-p*-toluenesulphonylglycine, and *n-p*-toluenesulphonyl-*dl*-α-alanine etc. (R″H = N-benzenesulphonyl-β-alanine), [Cu(Ph-big)(R_bH)], [Cu(Ph-big)($R_b^{\alpha}H$)], [Cu(Ph-big)($R_t^{\alpha}H$], [Cu(Ph-big)(R_bH)]. $0.5H_2O$, [Cu(Ph-big)(R_bH)] (R_bH_2 = *N*-benzenesulphonylglycine, $R_b^{\alpha}H_2$ = *N*-benzenesulphonyl-α-alanine, R_tH_2 = *N-p*-toluenesulphonylglycine) have been prepared and characterized [209]. The compounds are paramagnetic and show one absorption band in the region 17, 240-17, 540 cm^{-1}.

Compounds of types [Cu(Ph-big)(R_bH)] (λ_{max} = 580 nm; log ε = 1.79; λ_{mean} = 243 nm; log ε = 4.23), [Cu(Ph-big)(R_b^{α} H)] (λ_{max} = 580 nm; log ε = 1.73; λ_{mean} 242 nm; log ε = 4.19), [Cu (Ph-big)(R_tH)] (λ_{max} = 570 nm; log ε = 1.70; λ_{mean} = 244 nm; log ε = 4.20), [Cu(Ph-big)(R_t^{α} H)] (λ_{max} = 570 nm; log ε = 1.68; λ_{mean} = 244 nm; log ε = 4.19), [Cu(Ph-big)(R_bH)].$0.5H_2O$ (λ_{mean} = 580 nm; log ε = 1.78; λ_{max} = 243 nm; log ε = 4.20), [Cu(Ph-big) (R_tH)].$0.5H_2O$ (λ_{mean} = 580 nm; log ε = 1.78; λ_{max} = 243 nm log ε = 4.20), (Cu(Ph-big)(R_bH)] (λ_{mean} = 580 nm; log ε = 1.78; λ_{max} = 243 nm; log ε = 4.20) (R_bH_2 = *N*-benzenesulphonylglycine; R_b^{α} = *N*-benzenesulphonyl-α-alanine; R_tH_2 = *N-p*-toluenesulphonylglycine) have also been prepared and characterized. All these complexes show normal magnetic moments (1.80-1.92 B.M.) [209].

The first macrocyclic biguanide complexes, cyclo(1, 2)-dibiguanidinyl dibenzil, of metal(II) (M(II) = Cu(II), Ni(II)) were reported by Saha and Das [210]. The preparation starts with guanidine instead of biguanide as the amine fragment that condenses with benzil to form an intermediate Schiff base (Fig. 56), and two such Schiff base molecules under the template action of a

$$2H_2NC\,(=NH)\,NH_2 + O = C\,(C_6H_5)\,C\,(C_6H_5) = O \longrightarrow$$
$$[H_2NC(NH)N = C(C_6H_5)-]_2 + 2H_2O$$

Diguanidinyl benzil

Fig. 56

metal ion are made to cyclise together to generate biguanide bearing macro cyclic metal chelates (Fig. 57).

Copper(II) complexes of R-SalbigH (R = H, Me, Et, Ph, or 2-Et; and R-SalbigH = 2-salicylidinebiguanide or 5-substituted derivatives), which show subnormal magnetic moments, have been prepared by Poddar and Sarkar [211]. They were further investigated by Bera, Pramanik, and Sen [212], who

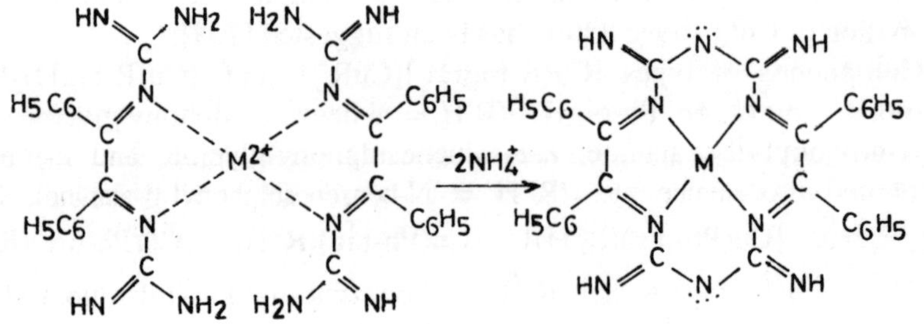

Fig. 57
Source : Indian Journal of Chemistry 32A (1993) 802

observed that with increase in the size of the substituent, the complexes change from a trimeric to a dimeric form.

The existence of a kinetically controlled first-wave in the polarographic reduction of the $[Cu(bigH)_2]^{2+}$ ion [213] was attributed to the reduction of the mono(biguanide) complex, formed by dissociation of the $[Cu(bigH)_2]^{2+}$ ion, which is facilitated by proton addition (Fig. 58) :

$$[Cu(bigH)_2]^{2+} + H^+ + 2H_2O \rightleftharpoons [Cu(H_2O)_2(bigH)]^{2+} + bigH_2^+ \quad (1)$$

Fig. 58

The formation of $bigH_2^+$ is favoured only at $pH < 7$. At still lower pH values (< 4.5) the mono complex again decomposes to give the $[Cu(H_2O)_6]^{2+}$ species. Polarograms at these pH values show the reduction wave of the Cu^{2+} ion along with that of $[Cu(H_2O)_2(bigH)]^{2+}$ ion; distinct $E_{1/2}$ values were obtained [214]. Thus, in the reduction of the $[Cu(bigH)_2]^{2+}$ ion the following equilibria are involved (Fig. 59) :

$$[Cu(bigH)_2]^{2+} + 2H_2O \underset{K^-}{\overset{K^+}{\rightleftharpoons}} [Cu(H_2O)_2(bigH)]^{2+} \; bigH \quad (2)$$

$$bigH + H^+ \overset{Kr}{\rightleftharpoons} bigH_2^+ \quad (3)$$

$$[Cu(H_2O)_2(bigH)]^{2+} + 2e \rightleftharpoons \text{Products} \quad (4)$$

Fig. 59

The stability constants of copper(II) biguanides were evaluated by Poddar, Poddar and Bhattacharya, [215], Doornbos [216], and Siefker and Wence [217]. Introduction of the heterocyclic ring enhances the basicity of the biguanide molecules, resulting in higher values of metal-ligand stability constants (log β) [215]. Banerjea found a relationship between log β_n and 10 Dq (~19000 cm^{-1}) for $[Cu(H_2O)_2(bigH)_2]^{2+}$ [218].

A paper chromatographic study [219] of square planar bis(biguanide) metal(II) chlorides [M(II) = Cu(II), Ni(II) and Pd(II)] in aqueous KCl-pyridine developer has shown that R_F values follow the order: Ni(II) > Cu(II) > Pd(II). Interestingly, R_F values increase with increasing alkyl substitution in the biguanide moiety of a given metal complex, but phenyl substitution lowers R_F values. A substantial amount of pyridine in the developer results in two spots of $[Cu(bigH)_2]Cl_2$, and the following equilibrium is suggested (Fig. 60).

$$[Cu(bigH)_2]Cl_2 + 4Py \rightleftharpoons [Cu(Py)_4]Cl_2 + 2bigH.$$

Fig. 60

Substitution of the hydrogen atom attached to the N^1 atom of bigH (I) considerably reduces the chances of forming H-bonded species with cellulose anions, which should lead to higher R_F values with increasing substitution. The increase of R_F values from bigH to Me-bigH and then to other alkyl biguanides (R-bigH) is also likely to be related to the inductive influence of the electron-releasing alkyl groups, whereby the overall charge of the complex cation may be reduced to some extent. This lowering of charge increases the R_F values. On the other hand, electron-withdrawing phenyl substitution leads to lower R_F values.

Dutta and Ray [220] noted that solubilities of the complexes in developers are believed to play a more significant role than the inductive effects of the substituent groups and H-bond-forming abilities of the ligand. Unfortunately, they could neither establish nor eliminate the following equilibrium (Fig. 61) :

$$[Cu(AAUH)_2]Cl_2 + 4Py \rightleftharpoons [Cu(Py)_4]Cl_2 + 2AAUH.$$

Fig. 61

Hota *et al* have prepared $[M(bigH)_2][Cu(acac)_2X]_2$ and $[Ni(R\text{-}bigH)_2]$ $[Cu(acac)Cl_2]$ (X = Cl, Br, or I) (M = Cu or Ni) (R = Me or Et) and characterized them by conductance, magnetic moments, thermal analyses, and, electronic and vibrational spectra [75] (Tables 8 and 9).

Râ̂y *et al* [221-223] have described the preparation of red and green (XXXIII) (Figs. 62, 63) forms of the copper complexes of 1-phenylbiguanide-*p*-sulphonic acid, [Cu (pbsa)$_2$], which they considered to be *cis* and *trans* isomers, respectively. However, on the basis of physical measurements (EPR; IR and absorption spectra), Wasson *et al* proved the red and green modifications to be the result of zwitterionic behaviour [224].

Copper(II) complexes of biguanides have been studied by EPR spectroscopy in water, in their polycrystalline form, and in the corresponding nickel(II)

Table 8 IR Spectral Data (cm^{-1}) for Some Metal Biguanide Complexes [75]

$[Cu(acac)_2Cl]_2$ $[Cu(bigH)_2]$	*$[Cu(acac)_2Br]$ $[Cu(bigH)_2]$*	*$[Cu(acac)_2Cl]_2$ $[Ni(bigH)_2]$*	*$[Cu(acac)_2Cl_2]$ $[Ni(Me\text{-}bigH)_2]$*	*$[Cu(acac)_2]$*	*$[Cu(bigH)_2]Cl_2$*	*$[Cu(big)_2].2H_2O$*	*bigH*	*Assignment*
3,460 s, 3,360 vs 3,225 w, 3160 w	3,450 s, 3,340 s 3,160 s	3,460 w, 3,350 m 3,150 ms	3,350 s, 3,250 w 3,145 w	– –	3,279 s	3,300 s	3,450 s 3,300 sh 3,200 sh	ν(N–H)
3,060 w, 2,920 w	3,040 w, 2,920 w	3,040 w, 2,910 w	3,020 w, 2,920 w	3,072 w, 2,987 w, 2,969 w, 2,920 w	2,941 w	–	–	ν(CH), ν(CH_3)
1,660 s	1,660 ms	1,660 s	1,650 s, 1,640 sh	–	1,653 m	1,660 vs	1,680 s 1,640 s	ν(N–C–N)
1,620 w	1,620 w	1,615 w	1,630 sh	–	–	–	–	δ(NH_2)
1,575 vs	1,575vs	1,570 w	1,560 ms	1,577 ms	–	–	–	ν(C–C–O)
1,530 s	1,530 s	1,530 s	1,525 s	–	1,541 m	1,580 vs 1,500 sh	1,523 sh	ν(N–C–N)
1,505 ms	1,505 w	1,505 m	1,505 m	1,529 m	–	–	–	ν(C–C–O)
1,430 sh	1,430 sh	1,430 sh	1,420 sh	1,461 sh	–	–	–	δ(C–H)+ δ(C–O)
1,390 vs	1,390 s	1,390 vs	1,390 vs	1,413 sh	–	–	–	δ(CH_3)
1,270 vs	1,270 s	1,270 s	1,260 vs	1,247 s	1,274 s	1,270 s		ν(C – CH_3) + ν(C–C)

$[Cu(acac)_2Cl]_2$ $[Cu(bigH)_2]$	$[Cu(acac)_2Br]$ $[Cu(bigH)_2]$	$[Cu(acac)_2Cl]_2$ $[Ni(bigH)_2]$	$[Cu(acac)_2Cl_2]$ $[Ni(Me\text{-}bigH)_2]$	$[Cu(acac)_2]$	$[Cu\text{-}(bigH)_2]\text{-}Cl_2$	$[Cu\text{-}(big)_2]\text{-}.2H_2O$	*bigH*	*Assignment*
1,020 vs	1,015 s	1,020 vs	1,010 ms, 990 w	1,019 ms				ν(CH_3)
930 vs	930 vs	930 vs	920 vs	936 ms	951	940		ν(C–C) + ν(C–O)
840 vs, 810 s	820 bw	835 ms 815 ms	830 ms, 800 ms	–	–	–	–	NH_2 wagging
780 s	770 ms	780 w	780 w	–	–	–	–	NH bending
680 s	680 s	680 m	680 w	684 m				ν(C–CH_3 + M-O+ring def)
650 ms	650 ms	640 ms	640 ms	653 ms				π
605 ms	605 ms	610 ms	610 ms	612 ms				ring def + ν(M–O)
520 ms	520 ms	{540 ms {520 ms	{540 w {495 s	–				ν(M–N)
435 ms	435 ms	430 ms	420 ms	451				ν(M–O)
290 w	290 w	290 w	290 w	290				–
240 w	–	245 w	240 mb	–				ν(M–Cl)

Table 9 Electronic Spectra of Some Copper(II) Complexes

Complex	*State*	*Absorption maxima* (cm^{-1})[a]
$[Cu(acac)_2Cl]_2\ [Cu(bigH)_2]$[b]	Nujol	10,990 sh – 13,330 sh – 16,000 – 18,870 sh
	Methanol	11,110 (35.5) – 13,160 sh (47.2) – 15,750 (66.5) – 18,870 sh (58.2)
$[Cu(acac)_2Br]_2\ [Cu(bigH)_2]$[b]	Nujol	10,810 sh – 13,330 sh – 16,130 – 18,870
	Methanol	10,990 sh (32.6) – 13,250 sh (46.5) – 15,870 (61.7) – 18,870 sh (54.6)
$[Cu(acac)_2I]_2\ [Cu(bigH)_2]$[b]	Nujol	11,110 sh – 13,330 sh – 16,260 – 18,870
	Methanol	11,110 sh (36.5) – 13,330 sh (43.5) – 16,130 (70.5) – 18,870 sh (51.2)
$[Cu(acac)_2Cl]_2\ [Ni(bigH)_2]$[b]	Nujol	10,990 sh – 13,340 sh – 16,000 – 21,739
	Methanol	11,111 sh (32.6) – 13,160 sh (41.2) – 15,870 (52.5) – 21,739 (18.5)
$[Cu(acac)_2Br]_2\ [Ni(bigH)_2]$[b]	Nujol	10,810 sh – 13,330 sh – 16,130 – 21,739
	Methanol	11,050 sh (34.6) – 13,330 (37.5) – 15,870 (51.2) – 21,739 (15.8)
$[Cu(acac)_2I_2]\ [Ni(bigH)_2]$[b]	Nujol	11,110 sh – 13,330 sh – 16,260 – 21,739
	Methanol	11,110 sh (32.6) – 13,330 sh (41.2) – 16,260 (52.5) – 21,739 (18.5)
$[Cu(acac)_2Cl_2]\ [Ni(Me\text{-}bigH)_2]$	Nujol	10,000 sh – 12,820 sh – 15,040 – 21,505
	Methanol	10,000 sh (28.2) – 12,820 sh (38.6) – 15,040 (54.2) – 21,505 (18.6)
$[Cu(bigH)_2]Cl_2.2H_2O$	Nujol[c]	20,000
	Single Crystal[c]	17,241 – 18,018 – 20,202
	Water[d]	19,230 (38.5)
	Methanol[b]	18,870 (43.1)
$[Cu(acac)_2Cl_2]\ [Ni(Et\text{-}bigH)_2]$[b]	Methanol	10,152 sh (34.5) – 12,820 sh (42.5)

[a]Values in parentheses indicate extinction coefficients [75].
[b]Ref. [139]
[c]Ref. [137]
[d]Ref. [140]

Table 10 g-Values and Hyperfine Structure Constants for Copper(II) Complexes

Complex	Medium	g_0	$g_{\parallel}$	$g_{\perp}$	$A_{\parallel}^{Cu}$	$A_{\perp}^{Cu}$	A_0^{Cu}	A_0^N	Ref.
						($\times 10^4$ cm^{-1})			
$[Cu(big)_2]$	Water	2.098					85	15.7	224
	Nickel(II)	–	2.169	2.062	208	24		16.0	224
$[Cu(bigH)_2]Br_2.2H_2O$	Nickel(II)	–	2.181	2.057	206	20		15.7	224
$[Cu(bigH)_2]Cl_2.2H_2O$	Nickel(II)	–	2.182	2.048	214	16	82	16.2	224, 226
$[Cu(Me\text{-}big)_2]$	Water	2.100	–	–	–	–	86.2	15.9	82
	Nickel(II)	–	2.172	2.060	209	24	–	16.0	82
$[Cu(Me\text{-}bigH)_2]Cl_2$	Water	2.099	–	–	–	–	85.9	15.8	82
	Nickel(II)	–	2.185	2.056	208	20	83.1	16.1	82
$[Cu(Et\text{-}big)_2]$	Water	2.098	–	–	–	–	87.1	15.8	82
	Nickel(II)	–	2.179	2.059	209	22	–	16.0	82
$[Cu(Et\text{-}bigH)_2]Cl_2$	Water	2.100	–	–	–	–	85.3	15.9	82
	Nickel(II)	–	2.189	2.064	206	22	–	15.7	82
$[Cu(Me_2\text{-}big)_2]$	Water	2.101	–	–	–	–	86.2	15.9	82
	Nickel(II)	–	2.182	2.057	208	20	–	15.7	82
$[Cu(Me_2\text{-}bigH)_2]Cl_2$	Water	2.098	–	–	–	–	85.8	15.8	82
	Nickel(II)	–	2.168	2.058	209	24	–	15.9	82
$[Cu(Et_2\text{-}big)_2]$	Water	2.099	–	–	–	–	85.6	15.1	82
	Nickel(II)		2.174	2.061	203	19	85.7	15.3	82
$[Cu(Et_2\text{-}bigH)_2]Cl_2$	Methanol	2.100	–	–	–	–	85.9	15.7	82
	Nickel(II)	–	2.169	2.054	210	21	–	15.6	82
$[Cu(Me_2\text{-}bigH)_2]Cl_2$	DMF	2.097	2.187	2.052	220	25	90	15	227

Complex	*Medium*	g_0	$g_\parallel$	$g_\perp$	$A_\parallel^{Cu}$	$A_\perp^{Cu}$	A_0^{Cu}	A_0^N	*Ref.*
						($\times 10^4$ cm^{-1})			
$[Cu(Me_2\text{-}big)_2].H_2O$	DMF	2.095	2.165	2.060	230	20	90	18	227
$[Cu(Me_2\text{-}bigH)_2]Br_2$	DMF	2.093	2.188	2.046	218	26	90	16	227
$[Cu(Me_2\text{-}bigH)_2](NO_3)_2$	DMF	2.093	2.193	2.043	217	26.5	90	15	227
$[Cu(Ph\text{-}bigH)_2]Cl_2$	Nickel(II)	–	2.158	2.052	203	38	–	18.3	233
$[Cu(Cl\text{-}Ph\text{-}bigH)_2]Cl_2$	DMF	2.105	2.193	2.061	207	33.5	91	15.3	229
$[Cu(Me_2\text{-}bigH)Cl_2]$	DMF	2.115	2.249	2.048	179	26	77	–	227
$[Cu(Me_2\text{-}bigH)(NO_3)_2]$	DMF	2.118	2.253	2.050	184	20.5	75	–	227
$[Cu(Me_2\text{-}bigH)(NCO)_2]$	DMF	2.118	2.245	2.054	184	20.5	75	14.6	227
$[Cu(Me_2\text{-}bigH)(NCS)_2]$	DMF	2.116	2.260	2.044	185	17	73	16	227
$[Cu(Me_2\text{-}bigH)(N_3)_2]$	DMF	2.103	2.226	2.041	194	29	84	–	227
$[Cu(Me_2\text{-}bigH)Br_2]$	DMF	2.116	2.267	2.041	174	27	76	–	227
$[Cu(morph\text{-}bigH)_2]Cl_2$	DMF	2.112	2.184	2.076	223	26.5	92	16.2	228
$[Cu(morph\text{-}bigH)_2]Br_2$	DMF	2.100	2.190	2.054	214	32.5	93	15.6	228
$[Cu(morph\text{-}bigH)_2](NO_3)_2$	DMF	2.098	2.188	2.054	209	30.5	90	16.0	228
$[Cu(morph\text{-}bigH)_2](NCO)_2$	DMF	2.094	2.167	2.058	226	23.5	91	16.0	228
$[Cu(morph\text{-}bigH)_2](OH)_2$	DMF	2.091	2.153	2.061	216	33	94	17.0	228
$[Cu(morph\text{-}bigH)Cl_2]$	DMF	2.120	2.253	2.054	167	29	75	17.5	228
$[Cu(morph\text{-}bigH)Br_2]$	DMF	2.119	2.278	2.040	171	27	75	17.5	228
$[Cu(morph\text{-}bigH)(NO_3)_2]$	DMF	2.125	2.269	2.053	175	29.5	78	16.5	228
$[Cu(morph\text{-}bigH)(NCO)_2]$	DMF	2.120	2.248	2.057	178	26.2	77	15.0	228
$[Cu(morph\text{-}bigH)(NCS)_2]$	DMF	2.126	2.261	2.059	185	15.5	72	16.2	228
$[Cu(morph\text{-}bigH)(N_3)_2]$	DMF	2.103	2.218	2.045	190	28	82	16.2	228

Complex	Medium	g_0	$g_\parallel$	$g_\perp$	$A_\parallel^{Cu}$	$A_\perp^{Cu}$	A_0^{Cu}	A_0^N	Ref.
						($\times 10^4$ cm^{-1})			
$[Cu(Cl\text{-}Ph\text{-}bigH)_2]Cl_2$	DMF	2.104	2.192	2.06	207	33.3	91	15.3	229
$[Cu(Cl\text{-}Ph\text{-}bigH)_2]Br_2$	DMF	2.103	2.187	2.06	211	31.5	91	15.4	229
$[Cu(Cl\text{-}Ph\text{-}bigH)_2](NO_3)_2$	DMF	2.103	2.181	2.06	212	38	96	15.3	229
$[Cu(Cl\text{-}Ph\text{-}bigH)_2](OH)_2$	DMF	2.103	2.182	2.063	209	28.7	89	15.1	229
$[Cu(Cl\text{-}Ph\text{-}bigH)Cl_2]$	DMF	2.122	2.259	2.054	176	40.5	86	–	229
$[Cu(Cl\text{-}Ph\text{-}bigH)Br_2]$	DMF	2.123	2.2577	1.056	176	34.5	82	–	229
$[Cu(Cl\text{-}Ph\text{-}bigH)(NO_3)_2]$	DMF	2.122	2.260	2.054	173	30.5	78	–	229
$[Cu(Cl\text{-}Ph\text{-}bigH)(NCO)_2]$	DMF	2.124	2.253	2.059	176	32	80	15.3	229
$[Cu(Cl\text{-}Ph\text{-}bigH)(NCS)_2]$	DMF	2.133	2.2609	2.069	176	27.5	77	15.3	229
$[Cu(Cl\text{-}Ph\text{-}bigH)(N_3)_2]$	DMF	2.106	2.225	2.047	187	29.5	82	–	229
$[Cu(amino\text{-}Et\text{-}bigH)(NCS)]SCN$	Single crystal	–	2.21	2.06	–	–	–	–	233
$[Cu(amino\text{-}Et\text{-}bigH)(C_2H_4N_4)]SO_4.H_2O$	Single crystal	–	2.19	2.048	–	–	–	–	233
$[Cu(pbsa)_2]$ (red)	Nickel(II)	–	2.175	2.057	201	26.9	–	15.7	201
$[Cu(pbsa)_2]$ (green)	Nickel(II)	–	2.155	2.046	216	25.8	–	15.3	201
$K_2[Cu(Pbsa)_2].H_2O$	Nickel(II)	–	2.149	2.063	209	28.9	–	14.4	201
$[Cu(Et_2\text{-}big)_2]$	DMF	2.090	2.150	2.060	218	35	96	16.9	233
$[Cu(Et_2\text{-}bigH)_2]Cl_2$	DMF	2.109	2.175	2.076	211	28	89	16.0	233
$[Cu(Et_2\text{-}bigH)_2]Br_2$	DMF	2.098	2.185	2.050	213	30	91	15.7	233
$[Cu(Et_2\text{-}bigH)_2](NO_3)_2$	DMF	2.097	2.179	2.056	208	29.5	89	15.8	233
$[Cu(Et_2\text{-}bigH)Cl_2]$	DMF	2.119	2.253	2.052	174	27	76	17.6	233
$[Cu(Et_2\text{-}bigH)Br_2]$	DMF	2.118	2.269	2.042	173	26	75	17.5	233
$[Cu(Et_2\text{-}bigH)(NO_3)_2]$	DMF	2.121	2.259	2.052	181	23.5	–	15.9	233

Complex	*Medium*	g_0	$g_{\parallel}$	$g_{\perp}$	$A_{\parallel}^{Cu}$	$A_{\perp}^{Cu}$	A_0^{Cu}	A_0^N	*Ref.*
						($\times 10^4$ cm^{-1})			
$[Cu(Et_2\text{-bigH})(N_3)_2]$	DMF	2.103	2.220	2.044	188	30.5	–	16.1	233
$[Cu(L_1)_2]$	Nickel(II) complex	–	2.099	2.059	208	23	–	15.8	232
	Water	2.099	–	–	–	–	85.9	15.7	232
$[Cu(L_1H)_2]Cl_2$	Nickel(II) complex	–	2.176	2.061	207	2	–	15.7	232
	Water	2.101	–	–	–	–	86.1	15.9	232
$[Cu(L_2)_2]$	Nickel(II) complex	–	2.173	2.057	209	24	–	16.0	232
	Water	2.101	–	–	–	–	86.2	15.9	232
$[Cu(L_3)_2]$	Nickel(II) complex	–	2.174	2.061	204	20	–	15.4	232
	Water	2.101	–	–	–	–	85.6	15.3	232
$[Cu(L_4)_2]$	Nickel(II) complex	–	2.173	2.063	203	19	–	15.3	232
	Water	2.099	–	–	–	–	85.6	15.2	232
$[Cu(L_5)_2]$	Nickel(II) complex	–	2.173	2.061	203	19	–	15.3	232
	Water	2.099	–	–	–	–	85.7	15.9	232
$[Cu(L_5H)_2]Cl_2$	Nickel(II) complex	–	2.168	2.058	210	25	–	16.0	232
	Water	2.098	–	–	–	–	86.1	15.9	232

L_1H = N^1-(2-hydroxyethyl) biguanide; L_2H = N^1-(2-hydroxypropyl) biguanide; L_3H = N^1-(3-hydroxypropyl) biguanide; L_4H = N^1-(2-methoxyethyl) biguanide; L_5H = N^1-(3-methoxypropyl) biguanide.

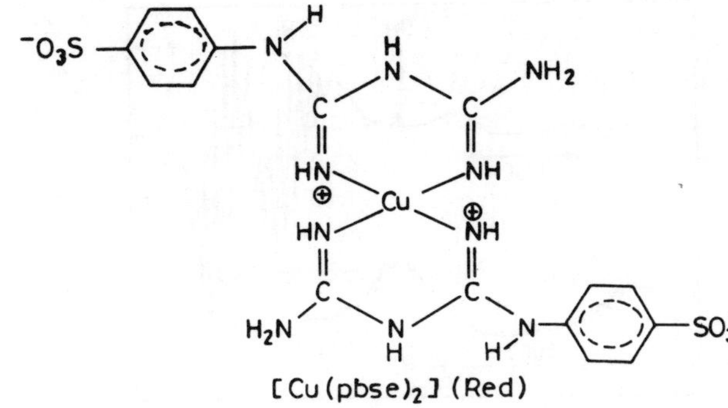

[Cu(pbse)$_2$] (Red)

Fig. 62

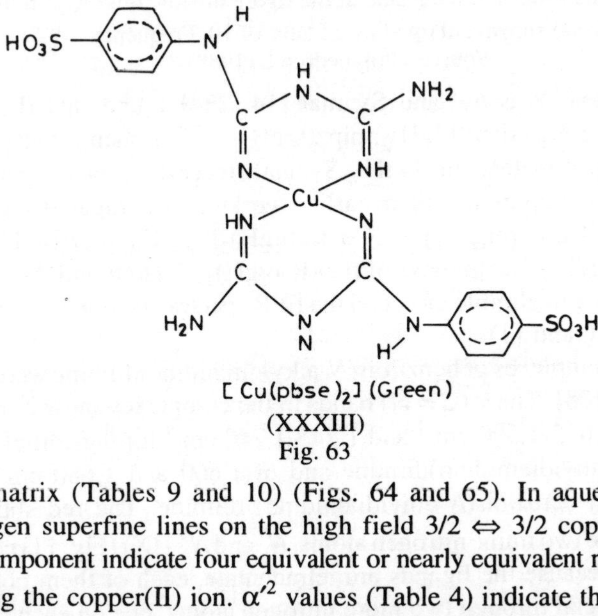

[Cu(pbse)$_2$] (Green)

(XXXIII)

Fig. 63

complex matrix (Tables 9 and 10) (Figs. 64 and 65). In aqueous solution, nine nitrogen superfine lines on the high field 3/2 ⇔ 3/2 copper hyperfine splitting component indicate four equivalent or nearly equivalent nitrogen atoms surrounding the copper(II) ion. α'^2 values (Table 4) indicate that, like bigH, AAUH, β-phthalocyanine, and tetraphenylporphine substituted biguanides and their protonated forms are equally strong field in nature [14, 82, 201, 224-232].

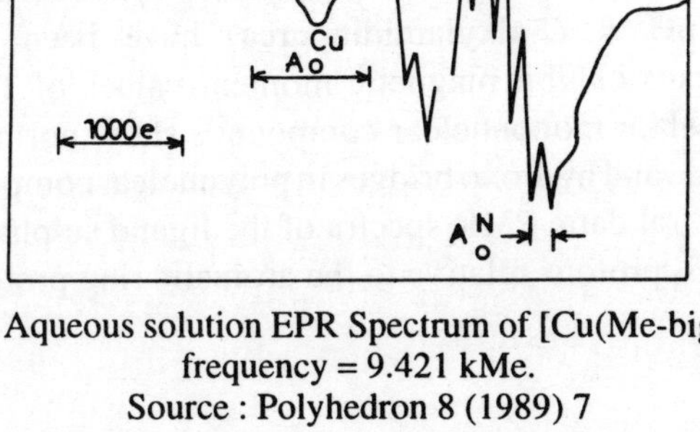

Fig. 64 Aqueous solution EPR Spectrum of [Cu(Me-bigH)$_2$]Cl$_2$ frequency = 9.421 kMe.
Source : Polyhedron 8 (1989) 7

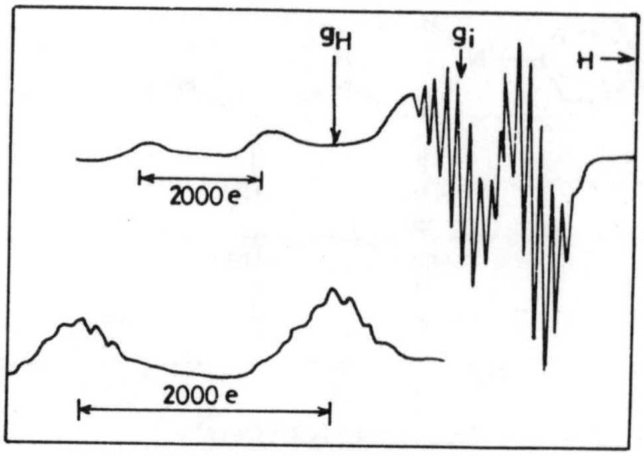

Fig. 65 EPR Spectrum of $[Cu(Me\text{-}bigH)_2]Cl_2$ in $[Ni(Me\text{-}bigH)_2]Cl_2$: Polycrystalline sample. In the lower left hand side of the figure the low field Al_{II}^{Cu} hyperfine components are shown enlarged by a factor of 10. Frequency = 9.518 kMe. Source : Polyhedron 8 (1989) 7

Rice, Stoklosa, Wasson, and Syamal [14, 234] found that the bridging groups in en$(bigH)_2$, trime$(bigH)_2$, pipz$(bigH)_2$ and *m*-phen$(bigH)_2$ influence the bonding parameters in bigH. Syamal recently repeated his earlier work [14, 230] suppressing their earlier works and compared the bonding parameters of $[Cu\ en(bigH)_2]^{2+}$, $[Cu\ tn(bigH)_2]^{2+}$, $[Cu\ pipz(bigH)_2]^{2+}$, and $[Cu\ m\text{-phen}(bigH)_2]^{2+}$ with those of $[Cu(R\text{-}bigH)_2]^{2+}$. He found that the bridging groups have a negligible effect on the EPR spectra and bonding parameters [230] (Tables 10 and 11).

Copper(II) complexes of benzilbis(*N*-alkyldiamidino)diimine were reported by Saha *et al* [108]. The ν (C = N) bands in the complexes show both red and blue shifts of 1,635-1,590 cm^{-1} and 1,745-1,740 cm^{-1} for the complexes with benzilbis(*N*-methyldiamidino)diimine and of 1,660 and 1,680 cm^{-1} for the complexes with benzilbis(*N*-ethyldiamidino) diimine. The red shift clearly indicates that the two imine nitrogen atoms, N^5 and $N^{5'}$ (IX) (Fig. 11) coordinate to the metal. Because the ligands are tetradentate, each of them coordinates further to the metal through two more nitrogen atoms, one on each biguanide part. The observed blue and red shifts of ν (C = N) further suggests that the N^5, $N^{5'}$, N^1, and $N^{1'}$ atoms coordinate to the metal. The electronic spectra of these complexes show onc asymmetric band at about 20,000-16,670 cm^{-1}, suggesting D_{4h} symmetry.

Polynuclear and mononuclear complexes of the types $Cu_3(Ln)_2(OH)_2$, $Cu_2(Ln)(L'n)(OH)$, and $Cu(LnH).2H_2O$ (LnH_2 = salicylidene-*O*-alkyl/benzyl-1-amidinourea) (L′nH = *O*-alkylamidinourea) have been reported. The polynuclear complexes exhibit magnetic moment values of 1.1-1.3 B.M. at room temperature, while mononuclear compounds show normal values. The presence of phenoxide and hydroxo bridges in polynuclear compounds has been suggested from spectral data. PMR spectra of the ligand sulphate show lower shielding of the $-NH_2$ protons relative to the aromatic ring proton [235].

Table 11 Orbital Reduction Factors and Covalency Parameters

Complex	*Medium*	*G*	$k_{\parallel}^2$	k_1^2	*k*	$\alpha_N'^2$	α_N^2	α_{Cu}^2	β_1^2	β^2	Δ_{Exy} (cm^{-1})	Δ_{Exz} (cm^{-1})	*Ref.*
$[Cu(big)_2]$	Nickel(II) complex	2.78	–	–	–	0.36	0.68	–	0.76	1.25	20,410	23,400	227, 228
$[Cu(Me\text{-}big)_2]$	Nickel(II) complex	2.93	0.52	0.82	–	0.36	0.72	0.81	0.72	1.14	20,410	23,400	82
$[Cu(Me\text{-}bigH)_2]Cl_2$	Nickel complex	3.35	0.55	0.76	–	0.35	0.73	0.82	0.76	1.05	20,410	23,400	82
	Water	–	–	–	–	0.36	0.72	–	0.73	1.00	20,410	23,400	82
$[Cu(Et\text{-}big)_2]$	Nickel(II) complex	3.10	0.54	0.83	–	0.36	0.72	0.82	0.75	1.16	20,410	23,400	82
$[Cu(Et\text{-}bigH)_2]Cl_2$	Nickel(H) complex	3.14	0.56	0.82	–	0.36	0.72	0.81	0.77	1.14	20,410	23,400	82
$[Cu(Me_2\text{-}big)_2]$	Nickel(II) complex	3.27	0.56	0.78	–	0.37	0.72	0.82	0.77	1.08	20,410	23,400	82
$[Cu(Me_2\text{-}bigH)_2]Cl_2$	Nickel(II) complex	2.96	0.51	0.79	–	0.36	0.72	0.81	0.71	1.10	20,410	23.400	82
$[Cu(Et_2\text{-}big)_2]$	Nickel(II) complex	2.92	0.53	0.83	–	0.36	0.72	0.80	0.74	1.15	20,410	23,400	82
$[Cu(Et_2\text{-}bigH)_2]Cl_2$	Nickel(II) complex	3.21	0.51	0.73	–	0.37	0.72	0.85	0.71	1.01	20,410	23,400	82
$[Cu(AMU)_2]$	Nickel(II) complex	3.26	0.50	0.71	–	0.35	0.69	0.81	0.70	1.05	20,410	23,530	82
$[Cu(Et_2\text{-}big)_2]$	DMF	2.55	0.46	0.82	0.84	0.34	0.74	0.82	0.61	1.11	20,350	23,400	233
$[Cu(Et_2\text{-}bigH)_2]Cl_2$	DMF	2.33	0.52	1.04	0.93	0.37	0.72	0.83	0.72	1.44	20,000	23,400	233
$[Cu(Et_2\text{-}bigH)_2]Br_2$	DMF	3.80	0.56	0.68	0.80	0.35	0.69	0.84	0.82	0.98	20,410	23,400	233
$[Cu(Et_2\text{-}bigH)_2](NO_3)_2$	DMF	3.28	0.55	0.76	0.83	0.34	0.74	0.82	0.74	1.03	20,410	23,400	233
$[Cu(Et_2\text{-}bigH)Cl_2]$	DMF	5.00	0.58	0.69	0.81	0.41	0.69	0.79	0.84	1.00	15,380	23,040	233
$[Cu(Et_2\text{-}bigH)Br_2]$	DMF	6.67	0.63	0.65	0.80	0.40	0.69	0.80	0.91	0.94	15,630	27,030	233

Complex	Medium	G	$k_{\parallel}^2$	$k_{\perp}^2$	k	$\alpha_N'^2$	α_N^2	α_{Cu}^2	β_1^2	β^2	Δ_{Exy} (cm^{-1})	Δ_{Exz} (cm^{-1})	Ref.
$[Cu(Et_2\text{-bigH})(NO_3)_2]$	DMF	5.14	0.60	0.69	0.81	0.40	0.69	0.82	0.87	1.00	15,630	23,040	233
$[Cu(Et_2\text{-bigH})(N_3)_2]$	DMF	5.19	0.53	0.68	0.79	0.36	0.72	0.80	0.73	0.94	16,150	27,030	233
$[Cu(L_1)_2]$	Nickel(II) complex	2.94	0.51	0.80	0.83	0.36	0.72	0.81	0.72	1.12	20,420	23,410	232
	Water	–	–	–	–	0.36	0.72	–	0.68	1.05	19,235	22,010	232
$[Cu(L_1H)_2]Cl_2$	Nickel(II) complex	2.94	0.53	0.83	0.85	0.37	0.72	0.81	0.74	1.15	20,410	23,400	232
	Water	–	–	–	–	0.37	0.72	–	0.70	1.08	19,230	22,000	232
$[Cu(L_2)_2]$	Nickel(II) complex	3.11	0.52	0.78	0.83	0.35	0.69	0.81	0.76	1.12	20,400	23,400	232
	Water	–	–	–	–	0.35	0.71	–	0.71	1.06	19,240	22,020	232
$[Cu(L_3)_2]$	Nickel(II) complex	2.91	0.53	0.83	0.85	0.35	0.69	0.80	0.76	1.20	20,410	22,440	232
	Water	–	–	–	–	0.35	0.69	–	0.71	1.06	19,230	22,000	232
$[Cu(L_4)_2]$	Nickel(II) complex	2.89	0.52	0.83	0.85	0.35	0.69	0.80	0.75	1.20	20,400	23,400	232
	Water	–	–	–	–	0.35	0.69	–	0.71	1.06	19,230	22,000	232
$[Cu(L_5)_2]$	Nickel(II) complex	2.89	0.52	0.83	0.85	0.35	0.69	0.80	0.75	1.20	20,415	20,450	232
$[Cu(L_5H)_2]Cl_2$	Nickel(II) complex	2.96	0.51	0.77	0.82	0.35	0.69	0.81	0.73	1.14	20,400	23,400	232
	Water	–	–	–	–	0.35	0.69	–	0.70	1.07	19,230	2,000	232

$L_1H = N^1$-(2-hydroxyethyl) biguanide;
$L_2H = N^1$-(2-hydroxypropyl) biguanide;
$L_3H = N^1$-(3-hydroxypropl) biguanide
$L_4H = N^1$-(2-methoxyethyl) biguanide;
$L_5H = N^1$-(3-methoxypropyl) biguanide

Copper(II) complexes of the type $CuL.H_2O$ (L = quadridentate Schiff base, bis(salicylidene)-*O*-alkyl-1-amidinourea) have been reported and characterized by elemental analysis, by IR and electronic spectra, and by magnetic moment, D.T.A., polarographic, and molecular weight measurements. The $-E_{1/2}$ values of the complexes are much less than those of the corresponding parent *O*-alkyl-1-amidinourea complexes, which may be due to the diminished ring current in the *O*-alkyl-1-amidinourea moiety in the present complexes. The stability of the complex with different Schiff bases with respect to the *O*-alkyl substituent are in the order: methyl > ethyl > propyl > butyl. The complexes show abnormal magnetic moment values (1.52-1.55 B.M.), which may be attributed to the presence of an oxo bridge as supported by IR spectral data [236].

The kinetics of the dissociation of $[Cu\ m\text{-phen}(bigH)_2]^{2+}$ and $[Cu\ en(bigH)_2]^{2+}$ in acidic media have been studied by Das *et al* [32]. Dissociation of the complexes occurs in two consecutive steps, the first being faster than the second. The infrared spectra, electronic spectra, and EPR parameters of some copper(II) complexes are shown in Tables 8, 9, 10, and 11.

(b) Copper(II) Heterochelates

Dipyridylmono(biguanide)copper(II) and related heterochelates

These compounds may be prepared by two different methods [237]. The first involves the addition of dipy (or *o*-phen) to blue $[Cu(H_2O)_2(bigH)]Cl_2$ at *p*H ~6. In this method the acidity of the solution is very important since formation of insoluble $[Cu(dipy)Cl_2]$ or $[Cu(o\text{-phen})Cl_2]$ is favoured at *p*H < 5 (Fig. 66):

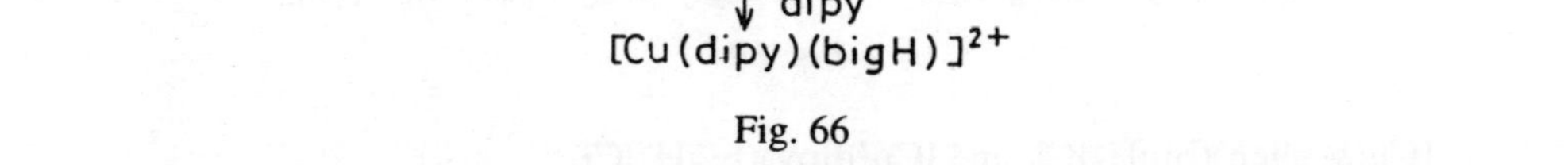

Fig. 66

In the second method the $[Cu\ (dipy)\ (bigH)]^{2+}$ ion is formed if bigH.HCl is slowly added to μ-diolbis (dipyridyl) dicopper (II) prepared *in situ* (Fig. 67).

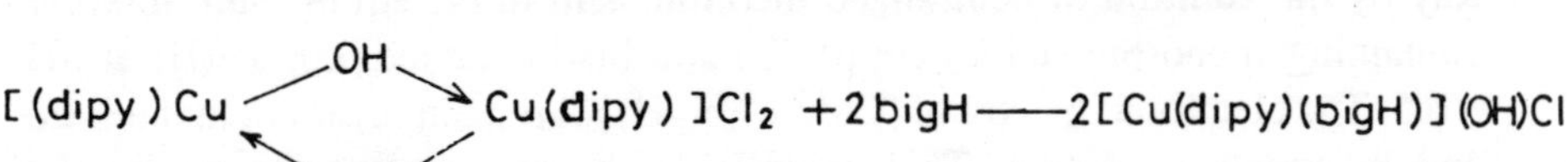

Fig. 67

The complexes are obtained as hydrates. Spectral studies indicate that these complexes have $[CuN_4O_2]$ chromophores (Table 12). The $[Cu(bigH)_2]^{2+}$ ion reacts with dipy and *o*-phen to form $[Cu(L)_2]^{2+}$ (L = dipy or *o*-phen) through intermediate formation of $[Cu(bigH)(L)]^{2+}$. Both of the steps in these reactions have been studied as well as the formation of $[Cu(dipy)(o\text{-phen})]^{2+}$ in the reaction of $[Cu(bigH)(dipy)]^{2+}$ with *o*-phen and of $[Cu(bigH)(o\text{-phen})]^{2+}$ with dipy.

In aqueous solution $[Cu(bigH)_2]^{2+}$ reacts with dipy and *o*-phen through intermediate formation of ternary complexes $[Cu(bigH)(L)]^{2+}$ and $[Cu(dipy)\text{–}(o\text{-phen})]^{2+}$ and binary complexes $[CuL_2]^{2+}$ (L = dipy or *o*-phen). The ratio of different steps has been followed in a borax buffer (pH = 8.0 ± 0.1) by stopped flow spectrophotometry [238, 239] (Fig. 68) (XXXIV).

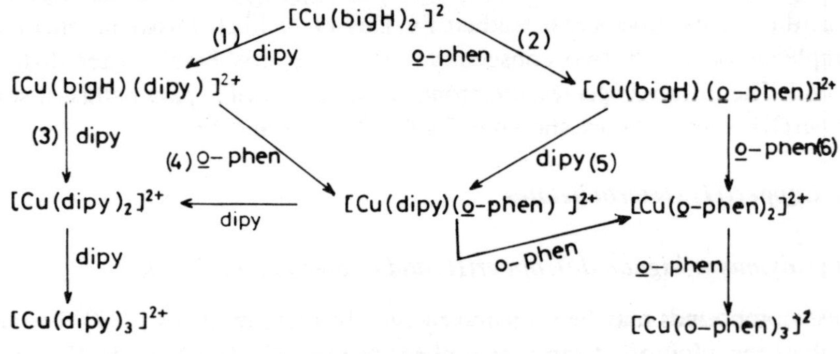

(XXXIV)
Fig. 68

In a paper chromatographic study of copper(II) heterochelate complexes using distilled water, aqueous potassium chloride, and an aqueous potassium chloride pyridine mixture as solvents, [Cu(gly-$\overline{\text{O}}$)(bigH)]Cl, [Cu(α-alan-$\overline{\text{O}}$)(bigH)]Cl, etc. give two spots, and the following equilibrium was suggested (Fig. 69).

$$2[Cu(gly\text{–}\overline{O})(bigH)]\,Cl \rightleftarrows [Cu(gly\text{–}\overline{O})_2] + [Cu(bigH)_2]Cl_2$$

Fig. 69

$[Cu(o\text{-phen})(bigH)]Cl_2$ and $[Cu(dipy)(bigH)]Cl_2$
provided single spots with the proper choice of developer [41, 240].

The picolinato mono(biguanide) copper(II) ion was prepared by Dutta and Ray by the addition of neutralized picolinic acid to the equilibrium mixture containing mono(biguanide) copper(II) and bis(biguanide) copper(II) at pH ~ 6.5. The electronic spectra of [Cu(pic)(bigH)]Cl remain invariant in the solid and in aqueous solution. This, coupled with the fact that the compound crystallizes from aqueous solution in anhydrous form, strongly suggests a square planar $[CuN_3O]$ chromophore for the complex [241].

Table 12 Colours and Electronic Spectra of Some Heterochelates [141, 237, 243, 244, 245]

Complex	*Colour*	*State*	*Absorption spectral band* (cm^{-1})	ε
[Cu(dipy)(bigH)]Cl_2.1.5H_2O	Royal blue	Aq. solution	16,950-15,630	63.5
	Royal blue	Solid	17,860-17,240	–
[Cu(dipy)(bigH)]I_2.H_2O	Deep blue	Aq. solution	16,660-15,630	66.0
	Sky-blue	Solid	16,390-15,630	–
	Light green with yellow tinge	Methanol solution	(i) 23,810-22,730 (ii) 13,510-12,820	(i) 203 (ii) 140
	Straw-yellow	Ethanol solution	(i) 23,260–22,730 (ii) 13,510–12,660	– (ii) 149
	Green	Ethylene glycol	(i) 24,390-23,260 (ii) 13,890-13,160	(i) 122.5 (ii) 99.0
	Green	DMSO	(i) 28,570 (ii) 15,870–14,490	(i) 493 (ii) 71.25
[Cu(*o*-phen)(bigH)]Cl_2.2H_2O	Deep blue	Aq. solution	16,950-15,630	55.50
[Cu(*o*-phen)(bigH)I_2.H_2O	Deep blue	Aq. solution	16,950-15,630	–
	Green	DMSO	(i) 28,570 (ii) 15,380-13,890	(i) 485 (ii) 75
	Yellowish green	Methanol solution	(i) 22,730-22,220 (ii) 13,330-12,820	(i) 242 (ii) 150
[Cu(dipy)(Me-bigH)]Cl_2.2H_2O	Royal blue	Aq. solution	16,400-15,630	68.00

Complex	*Colour*	*State*	*Absorption spectral band* (cm^{-1})	ε
[Cu(dipy)(Me-bigH)]I_2.0.5H_2O	Deep blue	Aq. solution	16,400-15,380	–
	Greenish-yellow	Methanol solution	(i) 22,730	–
			(ii) 13,510-12,820	–
[Cu(dipy)(Et-bigH)]Cl_2.3H_2O	Royal blue	Aq. solution	16,660-15,380	67.50
[Cu(dipy)(Et-bigH)]I_2	Deep blue	Aq. solution	16,950-15,380	–
	Greenish-yellow	Methanol solution	(i) 23,260	–
			(ii) 13,510–12,820	–
[Cu(IDA)(H_2O)(bigH)].0.5H_2O	Blue	Aq. solution	13,800-13,500	95
[Cu(pic)(bigH)]Cl	Deep blue	Aq. solution	16,700	42
	Deep blue	Solid	16,700	–
[Cu(nitrophen)(bigH)]Cl_2.0.5H_2O	Deep sky-blue	Aq. solution	16,670-15,620	31.65
	Light green	DMSO	14,290-13,330	69.00
	Deep sky-blue	Ethylene glycol	16,130-14,930	62.00
[Cu(nitrophen)(Me-bigH)]Cl_2.0.5H_2O	Deep sky-blue	Aq. solution	16,390-15,380	58.75
	Yellowish-green	DMSO	(i) 13,890-13,160	63.00
			(ii) 20,830-20,410 sh	43.50
	Light sky-blue	Ethylene glycol	14,930–14,080	63.00

Complex	*Colour*	*State*	*Absorption spectral band* (cm^{-1})	ε
$[Cu(nitrophen)(Et\text{-}bigH)]Cl_2.H_2O$	Deep sky-blue	Aq. solution	16,390-15,380	69.00
	Yellowish green	DMSO	(i) 13,890-13,160 (ii) 20,410-20,000 sh	63.00 45.00
	Light sky-blue	Ethylene glycol	14,930-14,080	63.00

ε = Molar extinction coefficient

$[Cu(IDA)(H_2O)(bigH)]0.5H_2O$ has been synthesized and characterized by Dutta, Banerjee, and De [242]. The formation constants of compounds of the type $[Cu(X)(Y)]Cl_2$ (X = bigH or R-bigH) (Y = dipy or *o*-phen) were also determined by Dutta, De and Syamal [237, 244], and these are of the order 10^8–10^9. Table 12 shows colours and electronic spectra of some copper(II) mixed chelates.

(c) Mono(biguanide)copper(II)Complexes

On addition of acid, red-violet bis(biguanide)copper(II) forms the deep blue diaqua mono(biguanide) copper(II), the formation of which is quantitative at *p*H 4.5-5.5 [207]. $[Cu(H_2O)_2(bigH)]Cl_2$ may also be prepared by mixing equimolar quantities of $[Cu(bigH)_2]Cl_2$ and $CuCl_2$ in water [1]. The colour change has been explained on the basis of stepwise dissociation of $[Cu(bigH)_2]Cl_2$ by H^+ ions (Fig. 70).

$$[Cu(bigH)_2]Cl_2 + HCl + 2H_2O \rightleftharpoons [Cu(H_2O)_2(bigH)]Cl_2 + bigH.HCl$$

$$\text{or } [Cu(H_2O)_2(bigH)]Cl_2 + HCl + 2H_2O \rightleftharpoons [Cu(H_2O)_4]Cl_2 + bigH.HCl$$

$$[Cu(bigH)_2]Cl_2 + [Cu(H_2O)_4]Cl_2 \rightleftharpoons 2[Cu(H_2O)_2(bigH)]Cl_2$$

Fig. 70

Compounds of type $[Cu(R\text{-}bigH)\,X_2]$ (R-bigH = Me_2-bigH, *p*-Cl-Ph-bigH, phenethylbigH, paludH, morph-bigH, piperidine-bigH) (X = Cl, Br, NO_3, NCO, NCS, NCSe or N_3) are also known and show the expected magnetic moments (μ_{eff} = 1.8 B.M.) at room temperature [192, 193, 197, 215, 227, 229]. Complexes of type $[Cu(AAA)(H_2O)]$ (AAA = glucosyl-big, glucosyl Me-big, glucosyl Et-big, glucosyl Me_2-big, glucosyl Et_2-big, glucosyl Ph-big) were prepared by the condensation of glucose and bigH/R-bigH with a copper(II) salt [245].

The zero field splitting parameter, $D'_{\|}$, and Cu-Cu distances of [Cu(aminoethybigH)(NCS)]SCN ($D'_{\|}$=0.048 cm^{-1}; 3.54 Å) [Cu(aminoethbig-H)$(C_2H_4N_4)]SO_4.H_2O$, ($D'_{\|}$ = 0.056 cm^{-1}, 3.44 Å) and were reported by Buluggiu, Dascola, Giori, and Varacca [246]. As $D_{\|}$ decreases, the Cu-Cu distances increase. The complexes are tetragonal, and IR data indicate that the coordination of thiocyanate and cyanate ligands occurs through the N atom.

(ii) Nickel(II) and (III) Complexes

A large number of nickel(II)bis(biguanide) and dibiguanide complexes are known; most are orange or yellow, but a few are red. These complexes resemble the corresponding copper(II) complexes in solubility, and all are decomposed by acids but unaffected by bases. The molar conductances of bis(biguanide) nickel(II) chloride in aqueous solution are in fair agreement with those for a bi-univalent electrolyte [194, 202, 215, 217, 253, 261]. In $[Ni(bigH)_2]Cl_2.2H_2O$ the water of crystallization is removed at ~ 125°C [208].

The diamagnetism and absorption band at 445 nm (22,472 cm^{-1}) in the solid state suggest a square planar configuration for $[Ni(bigH)_2]Cl_2$ [36, 79, 138, 139, 181, 182]. The polarized crystal spectrum of $[Ni(bigH)_2]Cl_2$ shows three bands at 430 nm (23,256 cm^{-1}), 460 nm (21,739 cm^{-1}), and 480 nm (20,833 cm^{-1}) [137, 142].

Solution spectra of other alkyl biguanide (R-bigH) nickel(II) complexes in different media have also been reported [31, 34, 143].

Of the two possible lowest energy transitions, $d_{xy} \rightarrow d_{x^2-y^2}$ ($^1A_{1g} \rightarrow {}^1A_{2g}$) and $d_{z^2} \rightarrow d_{x^2-y^2}$ ($^1A_{1g} \rightarrow {}^1B_{1g}$), for square planar d^8 complexes, if the latter involved the d_{z^2} orbital, that orbital being the one most exposed and therefore most strongly influenced by the solvent, a change of solvent should change the wavelength of the peak. Since the wavelength of the observed peak remains invariant on changing solvents, the lowest energy transition must be $d_{xy} \rightarrow d_{x^2-y^2}$ ($^1A_{1g} \rightarrow {}^1A_{2g}$), a conclusion that supports the $d_{xy} \rightarrow d_{z^2}$ order energy levels.

Nickel(II) complexes of *p*-anisidylbigH and Ph-bigH have been obtained in cream-coloured and red forms. These have different decomposition temperatures and electronic spectra [200]. The dissociation constant of bis (*o*-tolylbiguanide)nickel(II) complexes was found to be 2.67×10^{-7} [247]. The stability constants of some nickel(II)-biguanide complexes were measured by Poddar *et al* [215]. Heating $[Ni\ en(bigH)_2]Cl_2.H_2O$ in air to about 150°C causes the crystal to begin to shatter and become opaque, presumably because of loss of water of crystallization. Further heating causes darkening (260°C), and at about 330°C the residue becomes dark brown. At 350°C, it becomes black, presumably reflecting strong hydrogen bonding [181].

The $[Ni\ en(bigH)_2]^{2+}$ ion, having a large stability constant [1], is very stable in neutral solution, undergoes decomposition in acidic media (pH ~ 5), and follows a first order rate law [248] (Fig. 71).

$$[Ni\ en(bigH)_2]^{2+} + 2H^+ + nH_2O \longrightarrow [Ni(H_2O)_n]^{2+} + en(bigH_2^+)_2$$

Fig. 71

As the ligand is a strong base (*p*K = 13.08 ± 0.2), the addition of excess acid to a neutral solution of the complex shifts the equilibrium to the right. The following mechanism for the acid hydrolysis reaction is suggested (Fig. 72) (XXXV).

The true first order rate constants for unimolecular dissociation of one end of the ligand, producing a "half-bonded" intermediate, are represented by k'_1 and k''_1, and k' and k'' represent the equilibrium constants in the protonation equilibria.

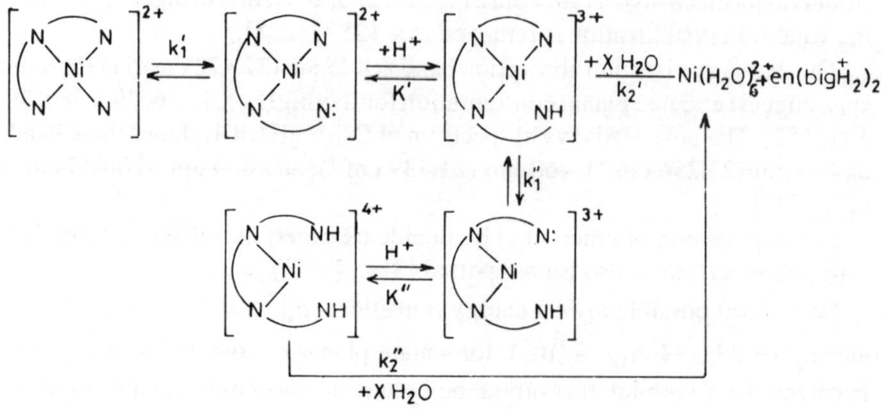

Fig. 72 (XXXV)

Banerjea [34] criticized the work of MacDonald [248]. According to Banerjea, the rate of dissociation of [Ni en(bigH)$_2$]$^{2+}$ cannot follow completely using *pH*-stat, since at the first stage, formation of "half-bonded" ethylenedibiguanide consumes protons, whereas the second step does not, as the en(bigH)$_2$ easily adds two protons in solution. However, the addition of two more protons is much less favoured thermodynamically as the log *K* values for these protonation processes indicate [248]. This is the reason why MacDonald observed a single-step reaction having a concurrent first and second order acid-dependent path. The kinetics of the dissociation of [Ni(bigH)$_2$]Cl$_2$ and [Ni en(bigH)$_2$]Cl$_2$ in acidic media forming the [Ni(aq)]$^{2+}$ and bigH$_2^+$ or [Ni(aq)]$^{2+}$ and en(bigH$_2^+$)$_2$ have been studied by Banerjea [34] using the stopped-flow technique. The reactions occur in two consecutive steps (the first being faster than the second), forming [Ni(H$_2$O)$_6$]$^{2+}$ and then en(bigH$_2^+$)$_2$ as the ultimate product. For each step the rate is acid-dependent, and the libility order [30] is as follows (Fig. 73).

$$[\text{Ni en(bigH)}_2]^{2+} > [\text{Ni(Ph-bigH)}_2]^{2+} > [\text{Ni(Me-bigH)}_2]^{2+} > [\text{Ni(bigH)}_2]^{2+}$$

Fig. 73

The kinetics of formation of square planar copper(II) and nickel(II) complexes of en(bigH)$_2$ has been studied spectrophotometrically in aqueous CH_3COOH–CH_3COONa buffers at 25-35°C. The observed rate constants for the formation reactions are independent of pH and of [CH_3COO^-] in the *p*H range used (3.6-4.8 for copper(II) and 5.0-5.8 for nickel(II)) [35].

From spectrophotometric and glass electrode measurements, MacDonald *et al* [249] redetermined the stability constant (β) and acid dissociation constant of [Ni en(bigH)$_2$]$^{2+}$ ion at 25°C: log β = 19.5 ± 0.2 (or β = 3 × 10^{19}); pK_1 = 12.075 ± 0.05; pK_2 = 13.08 ± 0.2. The reported values are β = 1.27 × 10^{15}; pK_1 = 11.8 and pK_2 = 11.3 [1, 124]. The main objection to the technique employed by Rây is that it is difficult to measure the acid dissociation constant of the conjugate acid form of the ligand, en$(bigH_2^+)_2$, with high precision, since it is an unusually weak acid. Therefore, its two successive acid dissociation constants are very close (Table 13).

The direct graphical method of data treatment used by Das Sarma [124] in his measurement of these acid dissociation constants is incapable of overcoming these difficulties. Furthermore, the $[Ni\ en(bigH)_2]^{2+}$ ion is kinetically inert. Therefore, aqua nickel(II) ion is slow to react with en$(bigH)_2$ when the *p*H of the solution is altered, and this takes sufficient time to reach equilibrium.

The tris(paludrine)nickel(II) complex has been prepared and characterized (μ_{eff} = 2.71 B.M. at room temperature) [144, 250]. The diamagnetic square planar mono(biguanide)nickel(II) complex has been synthesized by Spacu and others [144, 251]. Octahedral mixed ligand nickel(II) complexes of types $[Ni(en)(palud)_2].2H_2O$ and $[Ni(en)_2(palud)].2H_2O$ were reported and characterized by Spacu and Albescu [252, 253]. Another octahedral complex, $[Ni(H_2O)_2(palud)_2]Cl_2$, in which the two water molecules are located at the *trans* apical sites, was also reported [144, 251].

Recently, some square planar nickel(II) complexes of AAUH were reported by Ray, Bandyopadhyay, and Kauffman [7]. In donor solvents such as EtOH, DMF, DMSO and py, there is a shifting band position compared to water, unlike that observed in the square planar $[Ni\ en(bigH)_2]^{2+}$ ion. In the case of $[Ni\ en(bigH)_2]^{2+}$, the band position remains invariant, which means that the d_{-z^2} orbital which must receive the donor solvents in the *trans* apical sites is not the orbital immediately lower in energy than the $d_{-x^2-y^2}$ orbital. The order proposed for $[Ni\ en(bigH)_2]^{2+}$ is $d_{x^2-y^2} > d_{xy} > d_{z^2} > d_{xz}\ d_{yz}$. In the $[Ni(AAUH)_2]^{2+}$ complexes, the band shift indicates a tetragonality of the complex and a different *d*-orbital order, which is $d_{x^2-y^2} > d_{z^2} > d_{xy} > d_{xy} > d_{zx}$ with a 22,200 cm^{-1} band representing a $d_{z^2} \rightarrow d_{x^2-y^2}$ $(^1A_{1g} \rightarrow {}^1B_{1g})$ transition [7].

Oxidation of nickel(II)-substituted biguanide compounds with sodium chlorate in DMF and in basic media leads to the synthesis of mono-, bis- and tris nickel(III) complexes of the types $[Ni(paludH)_3]X_3$, $[Ni(H_2O)_2(paludH)_2]X_3$, $[NiCl_2(paludH)_2](OH)$ (X = Cl, Br, I, NCS, NCO, N_3, ClO_2, ClO_4, $\frac{1}{3}[Cr(NCS)_6]$ and $[Cr(NH_3)_2(NCS)_4]^-$). The complexes are paramagnetic (μ_{eff} = 1.66-1.78 B.M.), supporting the trivalent oxidation state for the metal. The bis- and tris (paludrine) nickel(III) complexes are octahedral. With NaH_2PO_2, N_2H_4 HCl and $SnCl_2$, the nickel(III) complexes are reduced to nickel(II) complexes [144, 250, 254]. A polarographic study of $[Ni\ (paludH)_3]^{3+}$ indicates one-electron reduction in DMF solvent (Fig. 74).

$$[Ni\ (paludH)_3]^{3+} + e \rightleftarrows [Ni\ (paludH)_3]^{2+}$$

Fig. 74

Other substituted biguanides (*o*-tolyl-bigH, Ph-bigH, *p*-Cl-PhbigH) of nickel(III) have been characterized [25]. Diamagnetic, octahedral complexes of the types [Ni(H_2O)(L)] (L = bases derived from glucose and bigH, Me-bigH, Et_2-bigH, Me_2-bigH/Ph-bigH) have been synthesized by Poddar [245]. Bubbling chlorine gas through [Ni(bigH)$_2$]Cl_2.2H_2O or [Ni(bigH)$_2$]Br_2 suspended in carbon tetrachloride yields [Ni Cl_2(bigH)$_2$]Cl or [Ni Br_2(bigH)$_2$]Br, respectively. In water the compounds are decomposed into the corresponding nickel (II) complexes. [Ni (bigH)$_2$$Cl_2$]X (X = Cl, F), [Ni(bigH)$_2$$Br_2$]Br, [Ni(AU)$_2$]Cl, [Ni(AMU)$_2$]$_2$$SO_4$.2$H_2O$, [Ni(AEU)$_2$]$_2$$SO_4$.2$H_2O$ [25] complexes are low spin $\left(s = \frac{1}{2}\right)$ and display polycrystalline EPR spectra that are rhombic. The probable structure is a basic grossly planar NiN_4 coordination sphere with axial ligation by Cl^-, Br^-, SO_4^{2-}, H_2O, etc. A $(d_{z^2})^1$ ground state is likely. The species are highly oxidising with E^0_{298} ~ 0.85 V in aqueous medium. Nickel(III) complexes of composition $[NiL_2(H_2O)_2]^{3+}$ (L = N^1, N^1-oxydiethylene-biguanide) and $[NiL_3]Cl_3$ (L = N^1-chlorophenyl-N^5-isopropylbiguanide) have also been reported [144, 250]. Polarographic studies of [NiCl_2(bigH)$_2$]Cl indicate that the one-electron reduction reaction is reversible with $E_{1/2}$ = – 0.40 volt. The average <g> value (2.18) supports a nickel(III) complex [25].

Complexes of the types [Ni(LH_2)]X_2 and [Ni(L′ H_2)]X_2 (X = Cl, Br, OH or CH_3COO) (LH_2 = benzilbis (*N*-methyldiamidino) diimine) (L′ H_2 = benzilbis(*N*-ethyldiamidino)diimine) are paramagnetic (μ_{eff} = 2.74 – 2.81 B.M.) and show three distinct bands at 24,100-23,250 cm^{-1}, 19,610-18,870 cm^{-1}, and 14,900-14,300 cm^{-1}, suggesting a D_{4h} local symmetry. The reduced magnetic moments are due to distortion from O_h symmetry. The spectral bands with decreasing wave number probably correspond to the transitions $^3A_{2g} \rightarrow {}^3T_{1g}$ (P), $^3A_{2g} \rightarrow {}^3T_{1g}$ (F), and $^3A_{2g} \rightarrow {}^3T_{2g}$ as in O_h symmetry. The middle band for the complexes of H_2L [H_2L = benzilbis (N-methyldiamidino) diimine] shows a shoulder at ~ 17,860 cm^{-1}, attributed to spin-orbit coupling, which mixes $^3T_{1g}$ (F) and ^1E_g states under the ligand field [108].

Thermogravimetric studies of [Ni(H_2O)$_2$(dichloro-bigH)$_2$]Cl_2 show an exothermic peak at 150°C with the loss of four chlorine atoms and the formation of [Ni(bigH)$_2$]Cl_2. A similar reaction is observed on heating the dibromo-bigH compound of nickel(II) [36]. A paper chromatographic study of square planar [M(R-bigH)$_2$]Cl_2 (M = Cu(II), Ni(II), Pd(II)) (R = H, Me, Et, Pr etc. and complexes by Dutta, Ray, and Kauffman [41] reported that the R_F values decrease in the order: Ni(II) > Cu(II) > Pd(II).

The polarographic behaviour of nickel(II)-biguanide/substituted biguanide complexes in aqueous potassium chloride containing gelatin has been reported [255]. The complexes undergo irreversible diffusion-controlled, two-electron transfer reduction at the d.m.e., except for $[Ni(bigH)_2]^{2+}$, which is irreversibly reduced. The $E_{1/2}$ values, rate constants, and activation energies of the complexes were found to be related to their corresponding average 10 Dq values.

Recently Ray and Kauffman [43] compared the movement of square planar metal(II) complexes on cellulose (paper) under identical conditions, with that of octahedral metal(III) and metal(II) complexes. The movement of octahedral cobalt(III) and nickel(II) complexes on paper was quite different from that of square planar nickel(II), copper(II), and palladium(II) complexes. For cationic octahedral complexes, outer sphere association occurs between the complex cations and the anions present in the developers, thereby reducing the overall charge of the complex cations and facilitating the movement of the cationic species along the negatively charged filter paper. For square planar complexes, direct coordination of the cellulose anion to the metal(II) ion occurs, thereby retaining the complex ion on the base line. However, with potassium chloride-pyridine developer, the cationic square planar complexes migrate readily and exhibit large spots.

(iii) Palladium(IV) and (II) Complexes

BigH and R-bigH compounds can stabilize palladium(IV) [256-258] to form complexes of the types $[PdX_2(bigH)_2]\ Y_2$ (X = Cl, Br, or OH) (Y = Cl, NO_3 or ClO_4), $[PdCl_2(py)_2(LH)]Cl_2$ (L = Ph-bigH, morph-bigH or paludH), $[PdCl_2Y_2(paludH)]$ (Y = Cl or NO_2), and $[Pd(paludH_2)_2][PdCl_6]$. Complexes of the type $[Pd(LH)_2]^{2+}$ and $[PdL_2]$ (LH = bigH or R-bigH) are light yellow, crystalline, diamagnetic and square planar. They are generally less stable than the corresponding nickel(II) complexes. The stability constant values for some palladium(II) complexes have been determined [216].

The crystal structure of $[Pd(bigH)_2](ClO_4)_2$ was determined by Nag and Guha [259]. In acidic medium (*p*H ~ 3.5), $[Pd(bigH)_2]Cl_2$ is converted to $[Pd(bigH)(H_2O)_2]\ Cl_2$ (λ_{max} = 345-350 nm; ε = 386), which has been isolated [260]. *Cis*-$[PdX_2L]$ (X = Cl, Br, N_3, NO_2) (L = big or R-big), $[PdL_2]\ [Pd(SCN)_4]$ and $[Pd(py)_2(R\text{-}bigH)](NO_2)_2$ (R-bigH = bigH or subst-bigH) have been prepared and characterized [135, 145, 257, 260, 262-268]. $[Pd(R\text{-}bigH)_2](SCN)_2$ (R = H, Me, Me_2 etc.) at *p*H ~ 2 undergo rearrangement to form a brown-red, insoluble, silky precipitate [96, 269] (Fig. 75).

$$2\left[Pd(R\text{-}bigH)_2\right](SCN)_2 + 2\,H^+ \longrightarrow [Pd\,(R\text{-}bigH)_2][Pd\,(SCN)_4]\downarrow + 2\,R\text{-}bigH_2^+$$

Fig. 75

Compounds of type $[Pd(N_3)_2(LH)]$ (LH = Me_2-bigH, morph-bigH, Ph-bigH, Cl-Ph-bigH, paludH, tolyl-bigH, naphtyl-bigH, or β-Ph(Et)-bigH) show a single frequency band at 2,060 cm^{-1}, supporting the existence of equivalent N_3^- groups [135]. This somewhat higher ν_{asy} (N_3) frequency is associated with the extraordinary stability of these azide complexes and strong Pd-N_3 bonds. The compounds are monomeric, have a *cis* configuration, and are not bridged via N_3^- groups [135]. The lowest energy bands at 29,000-30,000 cm^{-1} (more intense bands) are due to charge transfer $N_3 \rightarrow Pd$ ($\pi \rightarrow d_{x^2-y^2}$) and $^1A_{1g} \rightarrow {}^1A_{2g} + {}^1E_u$ (π). The second band at 39,000-40,000 cm^{-1} (shoulder) is due to charge transfer $L \rightarrow M$ ($e_u\, \sigma \rightarrow d_{x^2-y^2}$). The most pronounced band at 45,000 cm^{-1} is due to two different transitions, one corresponding to those in the conjugated –C=N–C=N– system, which in the free ligand appear at approximately 43,000 cm^{-1} and are shifted by coordination, and another corresponding to the azide group (44,000 cm^{-1}) and the alkyl azide group (46,000 cm^{-1}) [270].

The kinetic study of the acid hydrolysis of the [Pd en(bigH)$_2$]$^{2+}$ indicates that the complex undergoes decomposition in acidic media with the formation of [Pd(H$_2$O)$_n$]$^{2+}$ and en(bigH$_2^+$)$_2$ (Fig. 76).

$$[Pd\ en(bigH)_2]^{2+} + 2H^+ + nH_2O \longrightarrow [Pd(H_2O_n)]^{2+} + en(bigH_2^+)_2$$

Fig. 76

The hydrolysis occurs in two steps and follows a two-term rate law reflecting the contribution of singly protonated and doubly protonated species occurring at the reactive intermediates. Interestingly, the rate constant (k_2) is greater for the palladium(II) complex than for the nickel(II) complex by a factor of five, indicating that the protonation of the [Pd en(bigH)$_2$]$^{2+}$ ion occurs more easily than that of the [Ni en(bigH)$_2$]$^{2+}$ ion [249]. Stepwise dissociation of [Pd(bigH)$_2$]Cl_2 in hydrochloric acid-sodium chloride media was investigated by Banerjea and Banerjee [260]. At low acid concentration (0.01-0.1 *M* HCl), [Pd(bigH)$_2$]Cl_2 is totally transformed into [Pd(H$_2$O)$_2$(bigH)]Cl_2. At higher acid concentrations (0.15-0.4 *M* hydrochloric acid, [Pd(H$_2$O)$_2$(bigH)]Cl_2 again dissociates and forms the $[PdCl_4]^{2-}$ ion.

(iv) Platinum(IV) and (II)

Fusion of hexachloroplatinic(IV) acid with bigH.H_2SO_4 at ~ 160°C yields a yellow mass of [Pt(bigH)$_3$](SO_4)$_2$.2H_2O [256, 271]. The corresponding basic salt, [Pt(bigH)$_3$](OH)$_2SO_4$.4H_2O, is produced when the sulphate salt is treated with aqueous ammonia. Treatment of [Pt(bigH)$_3$](SO_4)$_2$ with concentrated sodium hydroxide solution yields yellows *cis*-[Pt(OH)$_2$(big)$_2$].2H_2O. A gray form of trans [Pt(OH)$_2$(big)$_2$].2H_2O was prepared by adding K_2PtCl_4, potassium hans chloroplatinate (IV), with an solution of bigH.H_2SO_4.

A heterochelate, [Pt(en)(big)$_2$]SO_4.2H_2O, is formed when yellow [Pt(OH)$_2$(big)$_2$].2H_2O and ethylenediamine are refluxed together. No such reaction occurs if grey [Pt(en)(big)$_2$].2H_2O and ethylenediamine are refluxed. On the basis of these chemical tests, the yellow compound was thought to be the *cis* isomer and the grey compound to be the corresponding *trans* isomer [256, 271]. *Trans-cis* isomerization did not occur even under boiling condition in this *trans*-platinum(IV) complex. However, in *trans*-cobalt(III) complexes with bigH, *trans-cis* isomerization takes place at steam bath temperature. All platinum(IV) biguanidle complexes are very stable to acids and alkalies.

The platinum complexes *trans*-$[PtX_2(paludH)_2]X_2$ (X = Cl or Br) were prepared by oxidation of $[Pt(paludH)_2]X_2$ with Cl_2 or Br_2 at *p*H 6-7. Treatment of paludH with $K_2[PtX_4]$ in neutral or slightly acidic medium yields $[PtX_2(paludH)_2]X_2$, but at *p*H ~ 3, $(paludH_2)_2[PtX_6]$ is formed [272]. With other biguanides, (LH = Me_2-bigH or morph-bigH) Pt(IV) compounds of composition $(LH)_2[PtX_6]$ (LH = Me_2-bigH or morph-bigH) and $[Pt X_2(L)_2]X_2$ (X = Cl, Br, I, or SCN) have been prepared.

Interestingly, in $[Pt(SCN)_2(Me\text{-}bigH)_2](SCN)_2$, SCN is bonded to Pt through S only, whereas in [Pt(NCS)(SCN) (morph-bigH)] $(SCN)_2$, SCN is linked to Pt both by N and S centers [273]. Oxidation of a concentrated solution of $[Pt(paludH)_2]Cl_2$ with sodium chlorite at 70°C yields octahedral $[Pt(PaludH)_3]Cl_4$ [273].

Digesting a platinum(II) chloride solution with free bigH base on a hot plate yields a wine-red solution, from which white crystals of $[Pt(bigH)_2]Cl_2$ separated. Gheorghiu prepared several platinum(II) complexes with Me_2-bigH, morph-bigH, phenethyl-bigH, or palud as ligands [272-274].

(v) Cobalt(I) Complexes

Das and De [22] prepared a cobalt (I)biguanide complex by reducing a cobalt-(II) or cobalt(III) biguanide complex with sodium borohydride in aqueous or nonaqueous media. The black compound thus formed is extractable in benzene or *n*-hexane. It is easily oxidized to cobalt (II) and possesses the catalytic property of reducing nitrate or nitrite to ammonia, acetate to acetaldehyde, benzoate to benzaldehyde and formate to formaldehyde [22].

(vi) Cobalt (II) Complexes

Rây and Ghosh first isolated solid, crystalline, greenish-yellow cobalt(II)-biguanide and dibiguanide complexes [1, 263]. Cobalt(II) chloride reacts with paludH in alcoholic media to form $[Co(H_2O)_2(paludH)_2]Cl_2$ [275]. Poddar, Ray, and Sarkar [203] prepared the cobalt (II) complex of biguanidebenzohydroxamic acid. Its subnormal magnetic moment (μ_{eff} = 0.8 B.M.) indicates some antiferromagnetic exchange, but the authors did not verify this by cryomagnetic or EPR studies. Ray also prepared $[Co(benzyl\text{-}big)_2]$ [100].

(vii) Cobalt(III) Complexes

(a) Tris(biguanide)Cobalt(III) Complexes

The original synthesis of tris(biguanide)cobalt(III) base and its salts involved an overcautious, cumbersome, and lengthy air oxidation of the insoluble bis (biguanide)cobalt(II) base in the presence of alkali biguanide [1, 276]. Dutta *et al* [277] prepared $[Co(bigH)_3](OH)_3$ and $[Co(bigH)_3]X_3$ (X = Cl, Br, I, $1/2SO_4^{2-}$

etc.) by slight modification (hydrogen peroxide oxidation) of the original procedure. Other tris(substituted biguanide)cobalt(III) bases and their corresponding salts have been reported (RbigH = cyclohexyl-bigH, 2-ethylhexyl-bigH, morph-bigH, piperidine-bigH, *n*-propyl-bigH, isopropyl-bigH or benzyl-bigH) [100, 194, 202, 215]. The formation constant of $[Co(bigH)_3]^{3+}$ ($\log K = 54$) is much greater than that of $[Co(en)_3]^{3+}$ ($\log K = 48.69$) [1, 19, 278]. Tris(biguanide)cobalt(III) chloride contains two water molecules which are lost at approximately 100°C [139]. Considerable delocalization of the ligand π-electrons through the d_π or p_π vacant orbitals of the metal is expected, as supported by an ultraviolet spectral study of $[Co(bigH)_3]Cl_3$, which shows two bands at 268 nm (ε = 4,013) and 190 nm (ε = 45,000). The first band is due to the $\pi-\pi^*$ transition of bigH, and the second one may be due to the aromatic ring containing the cobalt(III) and the bigH molecule [70, 130]. The second order paramagnetic term and electronic spectra of some cobalt(III)-biguanide complexes have been measured by Syamal [279]. The orbital reduction factors, ($K = 0.7 - 0.9$) of these complexes indicate their covalent nature.

The well-known rhombic twist mechanism proposed by Râу and Dutt for the trischelate $[Co(bigH)_3]^{3+}$ involves movement of two of the metal-ligand chelate rings in their respective planes, but in opposite directions. Movement of each of the two chelate rings through an angle of 45° in opposite directions leads to the transition state without any movement of the other (third) chelate ring; further movement of the two chelate rings through 45° leads to the mirror image form. Less energy is expended in the Ray and Dutt twist than in the Bailar twist [1, 27, 28]. Another type of study was conducted for *fac*-$[Co(Ph\text{-}bigH)_3]^{3+}$, in which a bond rupture mechanism was proposed regarding the intramolecular racemization process [64]. Interestingly, Kawaguchi *et al* [62, 63] found that optically active $[Co(am)(bigH)_2]^{2+}$ (am = aminoacidato ion) ions racemize more rapidly compared to $[Co(bigH)_3]^{3+}$ and $[Co(am)_3]$ complexes. The kinetics of dissociation in acidic media of $[M(R\text{-}bigH)_3]^{3+}$ (M = Co(III) or Cr(III)) (R = H, *n*-hexyl or Ph–) have been studied under comparable conditions, and the reaction may be represented by the equation (Fig. 77).

$$[M(bigH)_3]^{3+} + H^+ + 2H_2O \longrightarrow [M(H_2O)_2(bigH)_2]^{3+} + bigH_2^+$$

Fig. 77

These studies indicate that both for cobalt (III) and chromium (III) complexes of any particular series (tris or bis) the rate of substitution reactions obeys the order bigH > *n*-hexyl-bigH > Ph-bigH, [280, 281] which is also the order of decreasing basicity of the ligands and decreasing stability of the complexes. All these reactions are thought to occur through the formation of intermediate protonated species (Fig. 78).

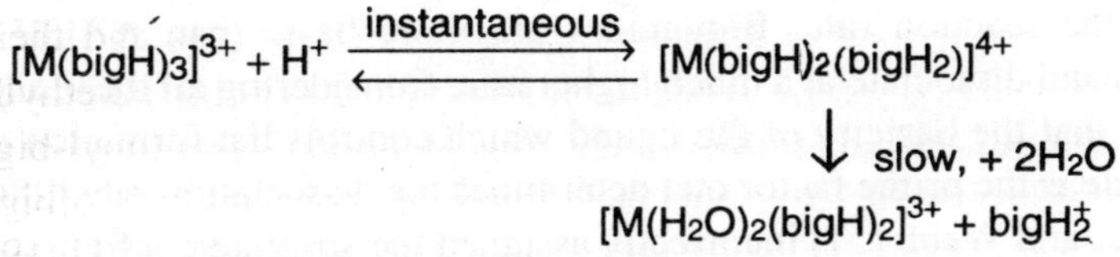

Fig. 78

The acid-catalyzed dissociation of metal-biguanide complexes was also reported by Banerjea *et al* [282]. The formation of $[Co(bigH)_3]^{3+}$ by the combination of bigH and *cis*-$[Co(H_2O)_2(bigH)_2]^{3+}$ in acidic media (pH = 2.0 – 8.2) has been investigated by Chakraborty and Sil [283]. In this pH range, $[Co(H_2O)_2(bigH)_2]^{3+}$ exists either as $[Co(H_2O)_2(bigH)_2]^{3+}$ or $[Co(H_2O)(OH)(bigH)_2]^{2+}$, and the ligand exists either as $bigH_2^+$ or ($bigH_3^{2+}$). The hydroxoaqua form is more reactive, and the reactions are entropy-controlled. Bond formation by the nucleophile occurs before completion of the bond-breaking stage in the transition state. At pH > 8, the bis complex may be partially transformed into its dihydroxo form [284].

The metal ion-catalyzed dissociation of $[M(bigH)_3]^{3+}$ (M = Co(III) or Cr(III)) in acidic media was investigated by Banerjee and Banerjea [285]; the value of k_M (specific rate constant for the metal-catalyzed path) increases linearly with $(Z - S)/r^2$ of the catalyzing metal ion, M^{2+} (M = Mn, Co, Ni, Cu, or Zn; Z = atomic number of M; S = Slater's screening constant; r = radius of M^{2+}), which is a measure of the complexing ability of M^{2+}. The catalytic activity of the different metal ions follows the order: Cu(II) > Ni(II) > Co(II) > Zn(II) > Mn(II). Banerjea noted that H^+ is a much better electrophile than M^{2+} ions for catalyzing the dissociation of $[M(bigH)_3]^{3+}$; the M^{2+} ion-catalyzed paths have higher enthalpy of activation than the proton-catalyzed path. This indicates significantly weaker binding of the catalyzing metal ion to the chelated biguanide, presumably as follows (Fig. 79) (XXXVI) :

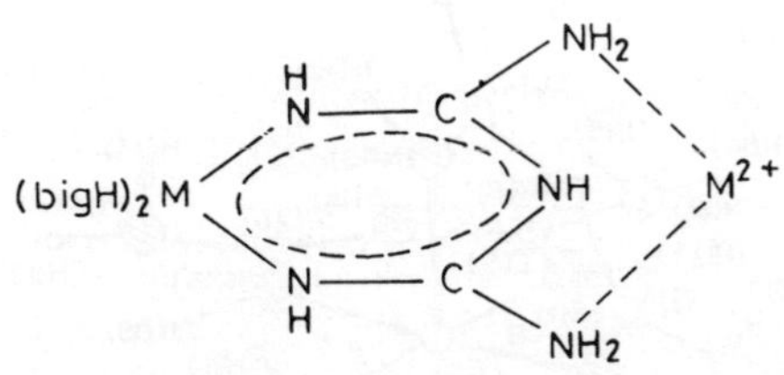

(M = Co(III) or Cr (III))

$M^{(2)}$ = Ni (II), Co (II), Zn (II), Mn (II)

(XXXVI)

Fig. 79

The kinetics of the acid-catalyzed dissociation of $[Co(AAUH)_3]^{3+}$ were investigated spectrophotometrically between 64°C and 78°C in perchloric acid. The complex dissociation is acid-dependent and follows a conjugate-acid mechanism. The rate constant depends on the basicity of the ligands.

Substitution of different alkyl groups on the side chain seems to have little effect on the reaction rate. Biguanides are more basic than AAUH form complexes and dissociate at a much higher rate. Considering all these aspects, it is likely that the basicity of the ligand which controls the formation of the intermediate is the prime factor that determines the dissociation rate [286].

Brubaker and Webb [57] incorrectly assigned the absolute configuration of $(-)_{589}-[Co(bigH)_3]^{3+}$ as Δ on the basis of the effect of PO_4^{3-} ion on the solution rotatory strengths in the visible region of the C.D. spectrum. Igi *et al* [58] have shown that the OH^- ion has the same effect and yields neutral, deprotonated $[Co\ (big)_3]$. These workers confirmed the validity of the visible UV-C.D. spectrum for $(-)_{589}-[Co(bigH)_3]^{3+}$ as reported by Michelsen [56], and they correctly suggested that the absolute configuration of $(-)_{589}-[Co(bigH)_3]^{3+}$ is Λ. In the ultraviolet region of the C.D. spectrum Michelsen [56] reported very intense bands for the cobalt and chromium complexes (same absolute configuration) at 247 (242 nm) and 218 (222 nm). The corresponding cobalt and chromium bands have opposite signs. Igi *et al* [58] suggested that these bands arise from $\pi-\pi^*$ transitions of the coordinated biguanide. Mason also measured the C.D. spectrum of $(-)_{589}-[Co(bigH)_3]^{3+}$ ion extended to lower wavelengths [287]. He assigned the negative 247 nm band to a charge transfer process, rather than an internal ligand transition, and he reported a new negative band at 197 nm. By X-ray techniques, Snow [30] also determined the absolute configuration of $(-)_{589}-[Co(bigH)_3]Cl_3.H_2O$ to be Λ. Fig. 80 shows the projection of the $[Co(big)_3]$ moiety. Coordination around cobalt is octahedral with three biguanides acting as chelating bidentate ligands and with Co–N bond distances (1.894-1.926 Å). Compairing fare well with those found in $[Co(bigH)_3]\ Cl_3.H_2O$ (1.902-1.922 Å). Each chelate ring is significantly now planar. Crystal packing (Fig. 81), together with the hydrogen bond contacts, is

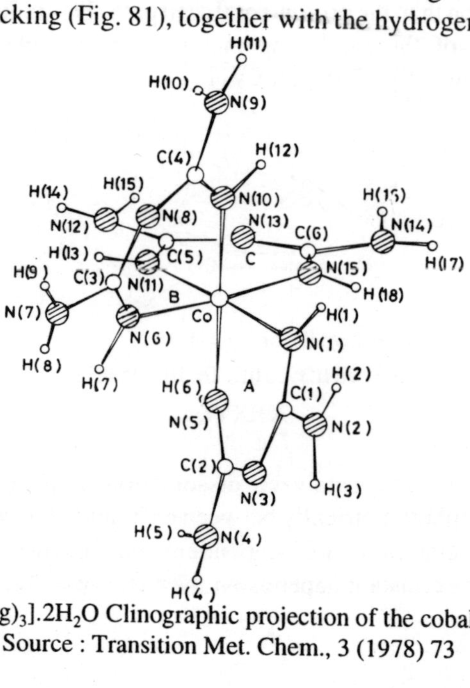

Fig. 80 $[Co\ (big)_3].2H_2O$ Clinographic projection of the cobalt environment
Source : Transition Met. Chem., 3 (1978) 73

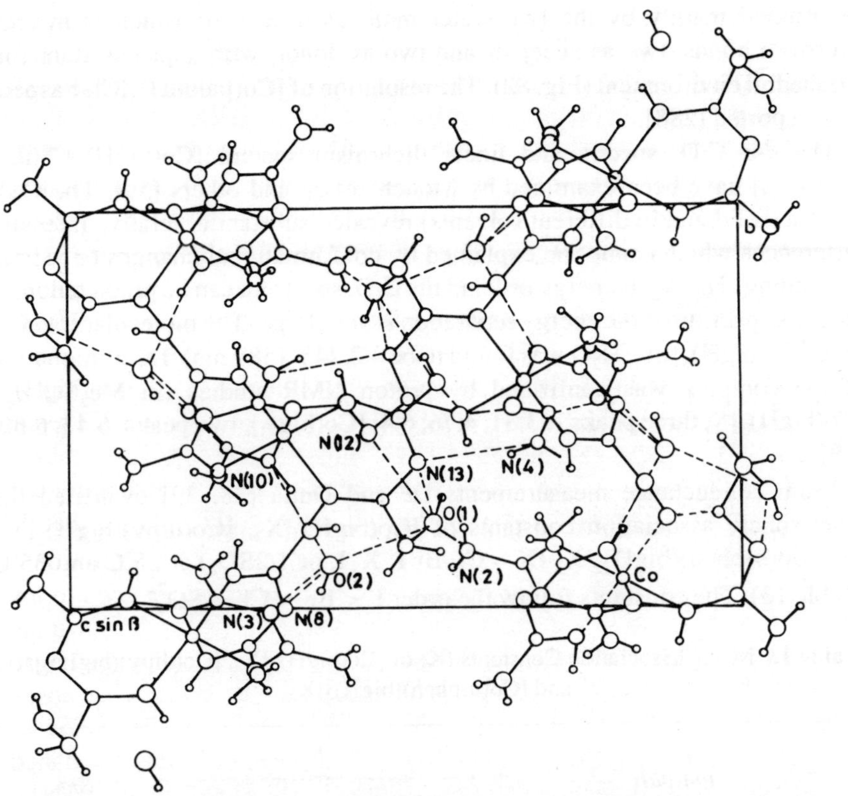

Fig. 81 $[Co(big)_3].2H_2O$: Projection of the structure along [100] showing crystal packing and hydrogen bonding
Source : Transition Met. Chem., 3 (1978) 73.

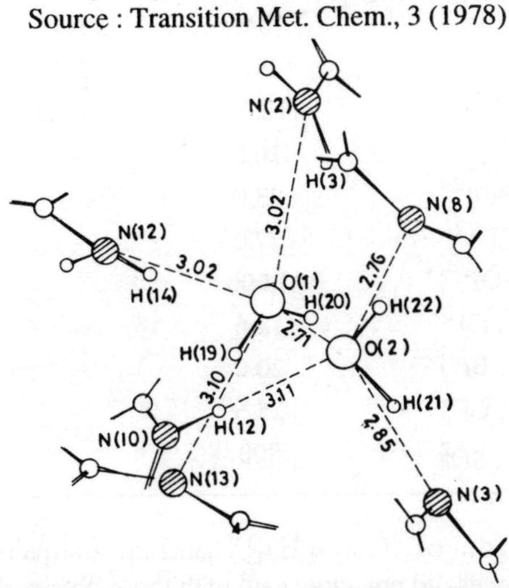

Fig. 82 $[Co(big)_3].2H_2O$: Hydrogen bonds involving the water molecules
Source : Transition Met., Chem., 3 (1978) 73.

determined mainly by the two water molecules, each of which is in four hydrogen bonds, two as acceptor and two as donor, with adjacent atoms, in tetrahedral environment (Fig. 82). The resolution of $[Co(paludH)_3]Cl_3$ has also been reported [288].

The *d-d* C.D. spectra and lineăr dichroism spectra $[Co(bigH)_3]^{3+}$ and $[Co(big)_3]$ have been examined by Moucharefieh and others [59]. The C.D. spectra (solid and in different solvents) revealed substantial relative intensity differences which cannot be explained by conformational changes or vicinal H-bonding. The A_2–E energy order in the C.D. spectrum can be predicted from the L.D. spectrum if the energy separation is very large. The molecular rotation of Λ-$[Co(bigH)_3]Br_3 .H_2O$ was found to be – 2,243° (589 nm). Deprontonation of the complex was confirmed by proton NMR studies (in Me_2SO–d_6 ; $[Co(bigH)_3]^{3+}$, three peaks, τ 3.31, 4.76, 6.4; $[Co(big)_3]$, two peaks, 6.43, 6.60) [59].

From conductance measurements De and Dutta [38, 39] evaluated the outer-sphere association constants of $[Co(bigH)_3]X_3$, $[Co(dipy)(bigH)_2]X_3$, and $[Co(o\text{-phen})(bigH)_2]X_3$ (X = Cl, Br I, X_3 I, or $1/2SO_4$) at 25°C and 35°C (Table 13). The constants follow the order $I^- < Br^- < Cl^- < SO_4^{2-}$.

Table 13 Mean Association Constants (K) of $[Co(bigH)_3]X_3$, $[Co(dipy)(bigH)_2]X_3$, and $[Co(o\text{-phen})(bigH)_2]X_3$

Ion-pair	*K*		*Ref.*
	25°C	*35°C*	
$\{[Co(bigH)_3]^{3+}, I^-\}^{2+}$	22	24	38
$\{[Co(bigH)_3]^{3+}, Br^-\}^{2+}$	34	37	38
$\{[Co(bigH)_3]^{3+}, Cl^-\}^{2+}$	55	62	38
$\{[Co(bigH)_3]^{3+}, SO_4^{2-}\}^+$	1,260	1,320	38
$\{[Co(dipy)(bigH)_2]^{3+}, I^-\}^{2+}$	16.6	–	39
$\{[Co(dipy)(bigH)_2]^{3+}, Br^-\}^{2+}$	23.0	–	39
$\{[Co(dipy)(bigH)_2]^{3+}, Cl^-\}^{2+}$	27.0	–	39
$\{[Co(dipy)(bigH)_2]^{3+}, SO_4^{2-}\}^+$	1,000	–	39
$\{[Co(o\text{-phen})(bigH)_2]^{3+}, I^-\}^{2+}$	12.6	–	39
$\{[Co(o\text{-phen})(bigH)_2]^{3+}, Br^-\}^{2+}$	20.0	–	39
$\{[Co(o\text{-phen})(bigH)_2]^{3+}, Cl^-\}^{2+}$	25.6	–	39
$\{[Co(o\text{-phen})(bigH)_2]^{3+}, SO_4^{2-}\}^+$	800	–	39

The approximate radii of $[Co(bigH)_3]^{3+}$ and the ion-pairs are shown in Table 14. Bjerrum distances did not agree well with those obtained from the sum of the anionic and cationic radii; the same value (5.2 Å) was obtained for the ion-pairs $\{[Co(bigH)_3]^{3+}, Cl^-\}^{2+}$, $\{[Co(bigH)_3, Br^-\}^{2+}$, and $\{[Co(bigH)_3]^{3+}, I^-\}^{2+}$.

Table 14 Approximate Radii of $[Co(bigH)_3]^{3+}$ and Ion-pairs

Cation/Anion	*Radius in Å from Stoke's law*	*Ion-pair*	*Radius in Å*	
			From Stoke's law	*From Bjerrum equation*
$[Co(bigH)_3]^{3+}$	4.03	$\{[Co(bigH)_3]^{3+}, Cl^-\}^{2+}$	5.23	4.28
Cl^-	1.20	$\{[Co(bigH)_3]^{3+}, Br^-\}^{2+}$	5.21	6.11
Br^-	1.18	$\{[Co(bigH)_3]^{3+}, I^-\}^{2+}$	5.22	7.38
I–	1.19	$\{[Co(bigH)_3]^{3+}, SO_4^{2-}\}^+$	6.32	5.035
SO_4^{2-}	2.29			

Eighteen cationic, neutral, and anionic cobalt(III) complexes were subjected to paper chromatography in aqueous developers containing varying concentrations of electrolytes such as potassium iodide, potassium chloride, potassium sulphate, sodium thiosulphate, and sodium potassium tartrate. In general, R_F values of all complexes belonging to a particular charge type (+ 3, + 2, or + 1) are lowest in KI, intermediate in KCl, and highest in K_2SO_4, sodium thiosulphate, and sodium potassium tartrate. This variation in R_F values conforms to the established order of outer-sphere association constants of complex cations with anions: $I^- < Cl^- << SO_4^{2-}$, $S_2O_3^{2-}$, $C_4H_4O_6^{2-}$ [43, 289]. Recently, Ray and others have shown that R_F values of a given complex cation in different developer solvents depend jointly on the surface tension of the electrolyte solution (expressed as the excess over the surface tension of pure water) and the anionic conductance of the electrolyte [44, 290]. Ray and Kauffman [45] used acidic mixed solvents (either formamide-methanol-acetic acid or DMF-methanol-perchloric acid) as the mobile phase and silica gel H as the stationary phase; they separated *cis* from *trans* isomers of some cobalt(III) complexes. The separation mechanism for these complexes is based on an ion-exchange process. Their finding that R_F values of *cis* isomers > R_F values of *trans* isomers is explained by the greater solubility of the *cis* isomer compared to that of the corresponding *trans* isomer. Ray and Kauffman [45] also observed that neutral and anionic cobalt(III) complexes move on a silica gel bed easily when distilled water is used as the mobile phase but that cationic complexes remained at the base line. These findings prompted them to seek a relationship between the R_F values and the charges of the complexes [289].

Polarographic reduction of $[Co(LH)_3]Cl_3$(LH = bigH or substituted bigH) in aqueous potassium chloride-containing gelatin has been reported [291]. The complexes are reduced in two steps (one electron and two electrons, respectively) at the d.m.e. The process is irreversible for all the complexes except for $[Co(bigH)_3]Cl_3$. The $E_{1/2}$ values, overall rate constant, and activation energies of the electrode processes have been found to be related to the average Dq values of the complexes.

(b) Cobalt(III) Heterochelates Containing Two Biguanides or One Dibiguanide and Another Bidentate Ligand

Heterochelates of type *o*-phen/dipy-bis(biguanide)cobalt(III) were prepared by treating *cis*-$[Co(NH_3)_2(bigH)_2](OH)_3$ with *o*-phen or dipy, followed by neutralization with 2 *N* hydrochloric acid or 2 *N* sulphuric acid to obtain the chloride or sulphate salts [146]. Similarly, complexes of type [Co(dipy or *o*-phen)en$(bigH)_2]^{3+}$ were prepared by treating dipy or *o*-phen with $[Co(NH_3)_2$-en$(bigH)_2]^{3+}$ and were characterized [292]. A comparison with the spectra of $[Co(bigH)_3]Cl_3$ ($\nu_1 = 21{,}100$ m^{-1}, $\nu_2 = 28{,}600$ cm^{-1}) suggested that the ligand field strengths of *o*-phen or dipy were slightly higher than that of bigH. The compounds were resolved into their enantiomers by means of their diastereoisomers with optically active anions. [Co(*o*-phen)$(bigH)_2]^{3+}$ showed a lower rotation than [Co(dipy)$(bigH)_2]^{3+}$, presumably because of the higher rigidity of *o*-phen compared to dipy.

Complexes containing two biguanides (or one en$(bigH)_2$) and one AAUH or *N*-phenylsalicylideneimine were prepared and characterized [146, 294, 295, 298]. Diastereoisomeric salts were isolated with sodium *d*-tartrate and *d*-camphor-10 sulphonate for resolution studies. Paper chromatographic studies of these complexes were carried out with distilled water and 1 *N* ethylamine as developer. All the mixed complexes containing AAUH provided single spots with reasonable R_F values [296].

Substitution of one bigH by Me-bigH or Et-bigH did not produce any asymmetric transformation of the second order with resnect to chloride *d*-tartrate diastereoisomers. Optical rotation values were indicative of greater asymmetry in the heterochelates. The diastereoisomers were stable in aqueous solution at about 30°C. When heated to 120°C, the solid compounds did not undergo racemization. Electronic spectra (Table 15) data for the mixed complexes as well as for $[Co(bigH)_3]Cl_3$ showed that the ligand field strengths of the substituted biguanides and AEUH were close to that of bigH.

Several salts of cobalt(III) mixed complexes containing acac or oxin and two big were also prepared and characterized [297]. Comparison of the absorption spectra showed that 8-hydroxyquinoline provides a stronger ligand field than acetylacetone (Table 15).

bphaH, acanH, picam, Me-bigH, Et-bigH and Ph-bigH were also used as a third participating ligand in the mixed complexes. The mixed chelates containing bpha and acan were bi-univalent electrolytes and exhibited two expected absorption bands in the electronic spectra, while the picam and other substituted biguanide complexes were tri-univalent. It is to be noted that $DMGH_2$ behaves as a bidentate dibasic acid rather than a bidentate monobasic acid in $[Co(DMG)(bigH)_2]^+$ and [Co(DMG)en$(bigH)_2]^+$ [301, 310].

Table 15 Electronic Spectra of Cobalt(III) Complexes in Aqueous Solution

Complex	Electronic spectra						
	Band II			Band I			
	λ_{max} (nm)	ν (cm^{-1})	ε	λ_{max} (nm)	ν (cm^{-1})	ε	Ref.
cis-$[Co(NH_3)_2(bigH)_2](SO_4)_{1.5}$				495	20,200	120	298
				490	20,400	–	–
trans-$[Co(NH_3)_2(bigH)_2](SO_4)_{1.5}$				495	20,200	80	298
cis-$[Co(OH)(H_2O)(bigH)_2]$*d*-tart				505	19,800	160	298
				510	19,600	–	–
$[Co(py)_2(bigH)_2](SO_4)_{1.5}.6H_2O$	370	27,030		505	19,800	–	299
$[Co(\gamma\text{-picoline})_2(bigH)_2](SO_4)_{1.5}.H_2O$				495	20,200	–	299
$[Co(OH)(\beta\text{-alan-}\overline{O})(bigH)_2](SO_4)_{0.5}.6H_2O$	375	26,670		480	20,830	–	299
$[Co(NO_2)_2(bigH)_2](SO_4)_{0.5}$	335	29,850		465	21,500	–	299
cis-$[Co(N_3)_2(bigH)_2]N_3$	296	33,783	316	508	19,685	168	300
trans-$[Co(N_3)_2(bigH)_2]N_3$	350	28,600	208	482	20,746	208	300
$[Co(Et\text{-}NH_2)(bigH)_2](SO_4)_{1.5}.4H_2O$	365	27,400		500	20,000		299
$[Co(o\text{-phen})(bigH)_2]Cl_3$	355	30,000	1,350	470	21,000	177	146
$[Co(dipy)(bigH)_2]Cl_3.3H_2O$	330	30,300	1,145	465	21,500	172	146
				470	21,300		
$[Co(en)(bigH)_2I_3$	349	28,600	155	480	20,900	165	301
$[Co(Me\text{-}bigH)(bigH)_2]Cl_3$	352	28,400	220	475	21,100	235	302

Complex	Electronic spectra						
	Band II			Band I			
	λ_{max} (nm)	ν (cm^{-1})	ε	λ_{max} (nm)	ν (cm^{-1})	ε	Ref.
$[Co(Et\text{-}bigH)(bigH)_2]Cl_3$	352	28,400	265	480	20,800	275	302
$[Co(Ph\text{-}bigH)(bigH)_2]Cl_3$	354	28,200	275	480	20,800	210	302
$[Co(acac)(bigH)_2]Cl_2$	340	29,400	1,600	500	20,000	177	302
$[Co(pic)(bigH)_2]Cl_2$		27,780		485	20,600	130	301
$[Co(oxin)(bigH)_2]Cl_2$	376	26,700	3,480	480	20,800	294	301
$[Co(DMG)(bigH)_2]Cl$				460	21,700	210	301
$[Co(dipic)(bigH)_2]Cl$				488	20,500	165	301
$[Co(IDA)(bigH)_2]Cl$	370	27,000	(sh)	488	20,500	163	301
$[Co(acan)(bigH)_2]Cl_2$	370	27,000	447	{505 500	19,800	167	297
$[Co(bpha)(bigH)_2]Cl_2$	370	27,000	440	510	20,000	207	297
$[Co(picam)(bigH)_2]Cl_3$				{475 946	19,600	177	297
$[Co(AMUH)(bigH)_2](SO_4)_{1.5}$	{349 355	{28,670 28,200	202	475	21,000	210	295
$[Co(AEUH)(bigH)_2]Cl_3$	355	28,200	210	475	21,100	225	295
$[Co(AMU)(bigH)_2]Cl_3$	335	29,850	200	480	20,800	180	295
$[Co(Sal\text{-}aniline)(bigH)_2]Cl_3$				500	20,000	120	303
$[Co(bigH)(trien)_2]Cl_3.H_2O$				487	20,530	142	303

Complex	Electronic spectra						
	Band II			Band I			
	λ_{max} (nm)	ν (cm^{-1})	ε	λ_{max} (nm)	ν (cm^{-1})	ε	Ref.
[Co(bigH)(trien)]I_3	353	28,300	180	485	20,600	130	303
[Co(bigH)(en)(trien)]	350	28,570	156	490	20,410	158	304
[Co(NO)(H_2O)(bigH)$_2$]Br_2	354	28,249	232	580	17,240	212	305
[Co(gly-$\overline{O}$)(bigH)$_2$]Cl_2.2H_2O	363	27,500	175	490	20,400	150	301
[Co(α-alan-$\overline{O}$)(bigH)$_2$]I_2	360	27,780	187	490	20,410	155	306
[Co(β-alan-$\overline{O}$)(bigH)$_2$]I_2				490	20,410		306
[Co(val-$\overline{O}$)(bigH)$_2$]I_2	360	27,780	199	485	20,620	161	306
[Co(en)$_2$(bigH)]ISO_4	350	28,600	132	480	20,800	158	319
				490	20,410		
[Co(leuc-$\overline{O}$)(bigH)$_2$]Cl_2	360	27,780	178	490	20,410	144	306
[Co(*dl*-methi-$\overline{O}$)(bigH)$_2$]I_2	358	27,900	214	490	20,410	165	307
[Co(hist-$\overline{O}$)(bigH)$_2$]I_2	–	–	–	475	21,100	235	306
[Co(bigH) en(bigH)$_2$]Cl_3	355	28,169	220	480	20,800	220	292
[Co(Me-bigH) en(bigH)$_2$]Cl_3.H_2O	355	28,169	220	480	20,800	225	293
[Co(Et-bigH) en(bigH)$_2$]Cl_3.H_2O	355	28,169	255	480	20,800	260	293
[Co(Pr^n-bigH) en(bigH)$_2$]Cl_3.H_2O	355	28,169	222	480	20,800	230	293
[Co(Pr^i-bigH) en(bigH)$_2$]Cl_3.H_2O	355	28,169	227	480	20,800	215	293
[Co(Ph-bigH) en(bigH)$_2$]Cl_3.2H_2O	355	28,200	218	480	20,800	227	293

Complex	Electronic spectra						
	Band II			Band I			
	λ_{max} (nm)	ν (cm^{-1})	ε	λ_{max} (nm)	ν (cm^{-1})	ε	Ref.
$[Co(en)\ en(bigH)_2](SO_4)_{1.5}.2.5H_2O$	350	28,600	167	480	20,800	173	293
$[Co(AMUH)\ en(bigH)_2]Cl_3$	355	28,200	208	475	21,100	220	293
$[Co(AEUH)\ en(bigH)_2]Cl_3$	355	28,200	210	475	21,100	230	293
$[Co(AP^nUH)\ en(bigH)_2]Cl_3$	355	28,200	206	475	21,100	205	293
$[Co(AP^iUH)\ en(bigH)_2]Cl_3$	355	28,200	173	475	21,100	197	293
$[Co(AB^nUH)\ en(bigH)_2]Cl_3$	355	28,200	171	475	21,100	193	293
$[Co(AB^iUH)\ en(bigH)_2]Cl_3$	355	28,200	165	475	21,100	190	293
$[Co(o\text{-phen})\ en(bigH)_2](SO_4)_{1.5}$	335	30,000	800	{480 496	{20,800 20,200	74	292
$[Co(dipy)\ en(bigH)_2](SO_4)_{1.5}$	355	28,200	220	{480 490	{20,800 20,400	220	292
$[Co(gly\text{-}\overline{O})\ en(bigH)_2]SO_4$	360	27,800	326	{515 530	{19,400 18,900	175	292
$[Co(\alpha\text{-alan-}\overline{O})\ en(bigH)_2]Cl_2.3H_2O$	360	27,780	322	490	20,410	173	308
$[Co(\beta\text{-alan-}\overline{O})\ en(bigH)_2]Cl_2.3H_2O$	360	27,780	320	490	20,410	171	308
$[Co(leuc\text{-}\overline{O})\ en(bigH)_2]Cl_2.5H_2O$	360	27,780	318	490	20,410	169	308
$[Co(val\text{-}\overline{O})\ en(bigH)_2]Cl_2.3H_2O$	360	27,780	326	490	20,410	178	308
$[Co(meth\text{-}\overline{O})\ en(bigH)_2]Cl_2$	355	28,200	309	490	20,410	177	309
$[Co(hist\text{-}\overline{O})\ en(bigH)_2]Cl_2.2H_2O$	–	–	–	475	21,100	175	308

Complex	*Electronic spectra*						
	Band II			*Band I*			
	λ_{max} *(nm)*	ν (cm^{-1})	ε	λ_{max} *(nm)*	ν (cm^{-1})	ε	*Ref.*
[Co(acac) en(bigH)$_2$]Cl$_2$.2H$_2$O	340	29,411	1,625	500	2,000	184	310
[Co(oxin) en(bigH)$_2$]Cl$_2$.2H$_2$O	375	26,666	3,481	480	20,833	294	310
[Co(pic) en(bigH)$_2$]Cl$_2$.2H$_2$O	–	–	–	485	20,618	135	310
[Co(gly-$\overline{O}$) en(bigH)$_2$]SO$_4$.1.5H$_2$O	360	27,777	326	515	19,417	175	310
[Co(dipic) en(bigH)$_2$]Cl.1.5H$_2$O	–	–	–	490	20,408	169	310
[Co(IDA) en(bigH)$_2$]Cl.1.5H$_2$O	370	27,027	–	490	20,408	166	310
[Co(DMG) en(bigH)$_2$]Cl.1.5H$_2$O	–	–	–	460	21,700	214	310
[Co(*l*-cys) en(bigH)$_2$]SO$_4$	–	–	–	475			*
[Co(*l*-cys) (bigH)$_2$]SO$_4$	338			470			*

*R.K. Ray and G.B. Kauffman, unpublished results

Substituted biguanides (R-bigH) and en react readily with $[Co(NH_3)_2en(bigH)_2](OH)SO_4$ to form red, crystalline mixed chelates of the type $[Co(R\text{-}bigH)en(bigH)_2]X_3.nH_2O$ (Tables 15 and 16). The complexes are 3 : 1 electrolytes in aqueous solution, diamagnetic, and they exhibit two ligand field bands, at 355 nm (band II) and 480 nm (band I) (Table 15) [293]. The optical rotation data for diastereoisomers and enantiomers of cobalt(III) mixed chelates are shown in Table 16.

(c) Cobalt(III) Heterochelates Containing Two Biguanides and One Potentially Tridentate Ligand

The potentially tridentate ligands used in mixed chelates of cobalt(III) containing two biguanides are iminodiacetic acid ($IDAH_2$) and dipicolinic acid ($dipicH_2$). These heterochelates were prepared from *cis*-$[Co(NH_3)_2(bigH)_2]$. Conductance

Table 16 Rotation Data for Diastereoisomers and Enantiomers of Cobalt(III) Heterochelates

Complex	*Specific rotation $[\alpha]_D$*	*Molar rotation $[M]_D$*	*Temperature (°C)*	*Ref.*
d-$[Co(o\text{-}phen)(bigH)_2]Cl(d\text{-}tart).5H_2O$	(+)383°	(+)2727°	36	146
l-$[Co(o\text{-}phen)(bigH)_2]Cl(d\text{-}tart).4H_2O$	(–)400°	(–)2776°	36	146
d-$[Co(o\text{-}phen)(bigH)_2](SO_4)_{1.5}.4H_2O$	(+)362°	(+)2371°	36	146
l-$[Co(o\text{-}phen)(bigH)_2](SO_4)_{1.5}.3.5H_2O$	(–)387°	(–)2500°	36	146
d-$[Co(dipy)(bigH)_2]Cl(d\text{-}tart).2H_2O$	(+)612°	(+)3880°	36	146
l-$[Co(dipy)(bigH)_2]Cl(d\text{-}tart).2.5H_2O$	(–)587°	(–)3774°	36	146
d-$[Co(dipy)(bigH)_2](SO_4)_{1.5}.2H_2O$	(+)644°	(+)3831°	36	146
l-$[Co(dipy)(bigH)_2](SO_4)_{1.5}.3.5H_2O$	(–)612°	(–)3806°	36	146
d-$[Co(bigH)_3]Cl(d\text{-}tart).5H_2O$	(+)340°	(+)2161°	31	146
l-$[Co(bigH)_3]Cl(d\text{-}tart).5H_2O$	(–)335°	(–)2129°	31	146
d- & *l*-$[Co(bigH)_3](SO_4)_{1.5}.3.5H_2O$	(±)362.5°	(±)2062°	31	146
l-$[Co(Me\text{-}bigH)(bigH)_2]Cl(d\text{-}tart).4.5H_2O$	(–)365°	(–)2328°	30	311
l-$[Co(Me\text{-}bigH)(bigH)_2](SO_4)_{1.5}.3.5H_2O$	(–)390°	(–)2266°	30	311
l-$[Co(Et\text{-}bigH)(bigH)_2]Cl(d\text{-}tart).4H_2O$	(–)375°	(–)2411°	30	311
l-$[Co(Et\text{-}bigH)(bigH)_2](SO_4)_{1.5}.3.5H_2O$	(–)385°	(–)2291°	30	311
d-$[Co(en)(bigH)_2](CS)_3.2.5H_2O$	(+)150°	(+)1590°	30	301
l-$[Co(en)(bigH)_2](CS)_3.2.5H_2O$	(–)150°	(–)1590°	30	301

Complex	*Specific rotation* $[\alpha]_D$	*Molar rotation* $[M]_D$	*Tempera-ture* (°C)	*Ref.*
d-$[Co(pic)(bigH)_2](CS)_2.3.5H_2O$	(+)250°	(+)2265°	31	301
l-$[Co(pic)(bigH)_2](CS)_2.3H_2O$	(–)250°	(–)2243°	31	301
d-$[Co(pic)(bigH)_2]I_2.2.5H_2O$	(+)200°	(+)1360°	28	301
l-$[Co(pic)(bigH)_2]I_2.1.5H_2O$	(–)350°	(–)2317°	28	301
l-$[Co(AMUH)(bigH)_2]\ I_3$	(–)250°	(–)1930°	28	5, 14
l-$[Co(AMUH)_2(bigH)](CS)_3$	(–)200°	(–)2174°	25	5
l-$[Co(AMUH)(bigH)_2]Cl(d\text{-tart}).4.5H_2O$	(–)375°	(–)2396°	25	5
l-$[Co(AEUH)(bigH)_2]Cl(d\text{-tart}).4.5H_2O$	(–)337°	(–)2197°	27	311
l-$[Co(AEUH)(bigH)_2](SO_4)_{1.5}.3H_2O$	(–)375°	(–)2197°	27	311
l-$[Co(Ap^i\text{-}UH)(bigH)_2]Cl(d\text{-tart}).4H_2O$	(–)325°	(–)2139°	25	5
l-$[Co(AP^i\text{-}UH)(bigH)_2](SO_4)_{1.5}.3H_2O$	(–)350°	(–)2104°	25	5
l-$[Co(AB^i\text{-}UH)(bigH)_2]Cl(d\text{-tart}).4.5H_2O$	(–)325°	(–)2213°	25	5
l-$[Co(AB^iUH)(bigH)_2](SO_4)_{1.5}.2H_2O$	(–)350°	(–)2090°	25	5

CSH = *d*-camphorsulphonic acid, $[M]_D = \dfrac{[\alpha]_d \times \text{Mol. Wt}}{100}$

data and equivalent weight determination proved them to be of univalent character. In $[Co(IDA)(bigH)_2]^+$ and $[Co(dipic)(bigH)_2]^+$ one ligand should exhibit flexidentate character to preserve the coordination number of cobalt(III) at six. The spectrum of $[Co(IDA)(bigH)_2]^+$ was similar to that of $[Co(gly\text{-}\overline{O})(bigH)_2]^{2+}$, while $[Co(dipic)(bigH)_2]^+$ and $[Co(pic)(bigH)_2]^{2+}$ were similar with respect to their electronic spectra (Table 15). Thus, the IDA^{2-} and $dipic^{2-}$ anions behaved like the gly-$\overline{O}$ and pic^- anions, respectively, although they are potentially tridentate ligands [301].

(d) Cobalt (III) Heterochelates Containing Two Bidentate Biguanides (or One Quadridentate Dibiguanide) and One Amino Acid

For syntheses of the mixed chelate $[Co(AB)(bigH)_2]^{2+}$ (ABH = amino acid), *cis*-$[Co(NH_3)_2(bigH)_2](OH)_3$ was found to be an ideal starting material. The complex base reacted readily with amino acids to liberate ammonia and provide the desired mixed chelates (Fig. 83).

$$[Co(NH_3)_2(bigH)_2]^{3+} \xrightarrow{AB^-} [Co(AB)(bigH)_2]^{2+} + 2NH_3\uparrow$$

Fig. 83

Orange-red [Co(*l*-hist-$\bar{O}$)(bigH)$_2$]X$_2$ (X = Br or I) shows one band at a higher energy (21,100 cm^{-1}) than the transition of the red-violet amino acid mixed chelates (20,400 cm^{-1}), indicating that the histidinate ion behaves as a NN bidentate donor and not a NO donor [301, 306]. The extreme similarities in the electronic spectra of [Co(AB)(bigH)$_2$]$^{2+}$ leave no doubt that all these mixed chelates have the same [CoN$_5$O] chromophore, i.e., methionine functions as a bidentate NO donor. The sulphur end of the methionine molecule remained unattached to cobalt(III) [306, 307, 309]. Recently Ray and Kauffman reported similar results while working on [Co(AB)(AAUH)$_2$]X$_2$ complexes [312].

During the acid-catalyzed dissociation of [Co(gly-$\bar{O}$)(bigH)$_2$]$^{2+}$ ion, two different isomeric (α- and β-) diaqua products were isolated. Up to 50°C the dissociation of only one biguanide takes place with the formation of the α-diaqua product. The α-diaqua form isomerizes to the β-form at higher temperatures. Direct formation of the β-form does not occur at any stage of the reaction. Isomerization of α-product takes place intramolecularly by two paths, one dependent and the other independent of the acid concentration. Both the isomerization paths involve a "bond-rupture process" [313, 314].

The formation of [Co(Ph-bigH)(bigH)$_2$]$^{3+}$ from [Co(H$_2$O)$_2$(bigH)$_2$]$^{3+}$ and its deprotonated form [Co(H$_2$O)(OH)(bigH)$_2$]$^{2+}$ with different protonated forms of Ph-bigH has been studied at *p*H 1.5-6.0. The monoprotonated form of the ligand takes part in the formation reaction. Ion-pair formation is completely absent, and multiple H-bonding is thought to be responsible for the association of the reactions. Significant bond formation occurs before bond breaking is complete in the transition state [283].

The acid-catalyzed aquation of the unsymmetrical complex [Co(pic)-(bigH)$_2$]$^{2+}$ at different temperatures has been studied by Chakraborty, Kar, and Ghosh [315]. As the reaction proceeds, one of the two biguanides was removed from the coordination sphere, and a dissociative mechanism is suggested. Substitution of water molecules from *cis*-[Co(H$_2$O)$_2$(bigH)$_2$]$^{3+}$ by bigH, Ph-bigH and picH is second order (first order each with respect to the complex and the ligand), and significant bond formation occurs before bond breaking is complete in the transition state. In these cases, ion-pair formation was totally absent [315-317]. The kinetics of the acid-catalyzed aquation of the heterochelate complex [Co(Ox)(bigH)$_2$]ClO$_4$(OxH$_2$ = oxalic acid) has been investigated at different temperatures. During the reaction, both the bigH moieties undergo substitution by the aqua ligands. Release of bigH moieties is acid-dependent, and bond breaking is important in the transition state [318].

Replacement of water molecules from *cis*-[Co(H$_2$O)$_2$(bigH)$_2$]$^{3+}$ by β-alan-OH and *l*-histOH has been investigated by Gangopadhyay and De [320]. The reaction rate is acid-dependent and involves an outer-sphere association between the two reacting species, followed by an associative interchange process where both bond breaking and bond making occur in a concerted manner. Generally, replacement of aqua ligands from aquacobalt(III) complexes follows dissociative pathways. Anation of [Co(H$_2$O)$_2$(bigH)$_2$]$^{3+}$ by histOH shows low $\Delta H^\ddagger$ and negative $\Delta S^\ddagger$ values, which indicates an associative mechanism. Biguanide itself is an unsaturated ligand and has a π-back-bonding tendency. It is likely that the π-bonding of the bis(chelate) ligand removes the electron cloud around the metal (cobalt or chromium) centers from the t_{2g} orbital, thereby increasing the tendency of these orbitals to accept the incoming ligand. Therefore, significant bond formation takes place before any bond breakage in the transition state. Such a mechanism is supported by low $\Delta H^\ddagger$ values—the energy of the bond breaking is partially compensated for by the energy released by bond formation by the incoming ligand [320].

The acid-catalyzed dissociation of the *trans*-$[Co(CN)_2(bigH)_2]^+$ ion has been studied by Chakraborty and Sii [321]. The following reaction takes place (Fig. 84).

$$[Co(CN)_2(bigH)_2]^+ + H^+ + 2H_2O \longrightarrow [Co(CN)_2(H_2O)_2(bigH)]^+ + bigH_2^+$$

Fig. 84

The electron donation by the CN^- ion to the cobalt(III) ion is reduced by the π-bond, making the cobalt atom substantially more positive than in $[Co(bigH)_3]^{3+}$. Therefore, the cobalt(III) ion in $[Co(en)(bigH)_2]^{3+}$ is the better acceptor of electrons, and the metal-biguanide bonds become shorter. Consequently, greater energy ($\Delta H^\ddagger$) is required for the rupture of the bond (Table 17). The $\Delta H\ddagger$ values for $[Co(CN)_2(bigH)_2]^+$ and $[Co(H_2O)_2(bigH)_2]^{3+}$ are equal, showing that the metal-ligand bonds are of equal strength. However, the reduced cationic charge of this complex favours the formation of the intermediate more than that of the diaqua complex with a positively charged proton, and the anation rate increases. The electron withdrawal by *trans*-CN^- ions lowers the electron density around the cobalt atom equally in all directions, and the bond formation by the incoming "water molecule" is not favoured until and unless the departure of the "bigH" takes place. Thus, bond breaking is the prime factor in the transition state [321].

Table 17 Activation Parameters of Various Substitution Reactions of Cobalt(III)-Biguanide Complexes

Complex	*Incoming group*	$\Delta H^\ddagger$ *(kcal/mole)*	$\Delta S^\ddagger$ *(e.u.)*	*Ref.*
$[Co(H_2O)_2(bigH)_2]^{3+}$	Ph-$bigH_2^+$	13.8	−20.8	283
$[Co(H_2O)_2(bigH)_2]^{3+}$	$bigH_2^+$	20.2	−3.5 ± 1.3	322
$[Co(H_2O)(OH)(bigH)_2]^{2+}$	Ph-$bigH_2^+$	14.0	−17.2	283
$[Co(H_2O)(OH)(bigH)_2]^{2+}$	$bigH_2^+$	19.8 ± 1.1	−3.4 ± 1.6	322
$[Co(NH_3)_2(bigH)_2]^{3+}$	*o*-phen	31.2	23.8	323
$[Co(NH_3)_2(bigH)_2]^{3+}$	dipy	30.6	22.2	323
$[Co(H_2O)_2(en)_2]^{3+}$	*o*-phen	31.1	10.1	324

Complex	Incoming group	$\Delta H^{\ddagger}$ (kcal/mole)	$\Delta S^{\ddagger}$ (e.u.)	Ref.
$[Co(H_2O)_2(en)_2]^{3+}$	dipy	30.7	8.5	325
$[Co(H_2O)_2(bigH)_2]^{3+}$	histOH	10.4	–	320
$[Co(H_2O)_2(bigH)_2]^{3+}$	Ph-alanOH	11.0	–	320
$[Co(H_2O)_2(en)_2]^{3+}$	H_2O	28.9	+15.1	326
$[Co(H_2O)_2(en)_2]^{3+}$	OxH_2	24.8	1.5	327
$[Co(H_2O)(NO_2)(DMG)_2]$	Br^-	22.5	–	328
$[Co(H_2O)(I)(DMG)_2]$	Br^-	17.8	–14.0	328
$[Co(H_2O)_2(DMG)_2]$	Cl^-, Br^-, I^-	16.19	–	329
$[Co(bigH)_3]^{3+}$	H_3O^+	11.1	–31.5	281
$[Co(H_2O)_2(bigH)_2]^{3+}$	H_3O^+	15.2	–31.6	316
$[Co(acac)(bigH)_2]^{2+}$	H_3O^+	15.0	–21	316
$[Co(Ph\text{-}bigH)_3]^{3+}$	H_3O^+	14.3	–25.1	281
$[Co(hex\text{-}bigH)_3]^{3+}$	H_3O^+	11.1	–33.4	281
$[Co(AMUH)_3]^{3+}$	H_3O^+	17.0 ± 1.0	– 41 ± 4	286
$[Co(AEUH)_3]^{3+}$	H_3O^+	19.2 ± 1.1	–33.7 ± 4	286
$[Co(AB^nUH)_3]^{3+}$	H_3O^+	28.5 ± 3.0	–7.5 ± 9	286
$[Co(en)_2(bigH)]^{3+}$	H_3O^+	46 ± 4.6	48 ± 8.2	330
$[Co(en)(bigH)_2]^{3+}$	H_3O^+	20.2 ± 1.1	–18.4 ± 2.1	330
$[Co(CN)_2(bigH)_2]^{+}$	H_3O^+	15.6	–24	321
$[Co(gly\text{-}\overline{O})(bigH)_2]^{2+}$	H_3O^+	17.4	–22.4	330

Quasi-aromatic Ph-bigH and acetylacetonate rings are strong metal-to-ligand π-electron acceptors, and their presence gives a very high rate of aquation. Five-membered rings of glycinate and ethylenediamine are unable to act as π-acceptors and do not enhance the rate of aquation of a biguanide complex. In the diaqua complex, the weak donor properties of the aqua ligands apparently make the cobalt atom less negative than in $[Co(bigH)_3]^{3+}$. The greater positive charge on the metal atom in this complex makes the cobalt atom a better acceptor of ligand electrons and thus stabilizes the big ligand, giving a slower rate for its dissociation. Therefore, the more the increase in effective positive charge on the metal atom due to greater electron withdrawal on weaker electron donation by the nonlabile ligand or ligands, the more will be the rate of dissociation of bigH, indicating a decrease in the kinetic stability of the chelate [330].

The rate of the substitution reactions follows the order of the basicity of the ligand, i.e., en (1.0) < DMG (10.0) < bigH (10^3), which suggests that bis(biguanide)cobalt(III) has some class (b) character [283].

(e) Cobalt(III) Chelates Containing Two Bidentate Biguanides (or One Quadridentate Dibiguanide) and Two Same or Different Monodentate Ligands

The original tedious procedure for the synthesis of *cis*-$[Co(NH_3)_2(bigH)_2]^{3+}$ involved air oxidation of freshly precipitated bis(biguanide)cobalt(II) in the presence of concentrated ammonia, followed by filtration and neutralization to provide the sulphate salt. Aeration for an hour did not ensure even 40% oxidation, and crystallization of the sulphate often required several days' standing in the refrigerator. Dutta and Sarkar improved this method by using hydrogen peroxide as the oxidizing agent. The base $[Co(NH_3)_2(bigH)_2](OH)_3.nH_2O$ or its salts can be prepared in between three and four hours and are widely used as starting materials for various studies. Both the *cis* and *trans* isomers show one absorption band at 20,200 cm^{-1}, but the *cis* isomer shows a higher extinction coefficient ($\varepsilon = 120$) than the *trans* isomer ($\varepsilon = 80$) [14].

Treatment of $[Co(NH_3)_2(bigH)_2]Cl_3$ with excess sodium *d*-tartrate yielded red crystals of the chloride-*d*-tartrate, which, on hydrolysis in aqueous medium produced violet crystals of (±)-*cis*-$[Co(OH)(H_2O)(bigH)_2]$ *d*-tart ($[\alpha]_D$ = (+)443° and $[M]_D^{22^\circ C}$ = (+) 2197°). *Trans*-$[Co(NH_3)_2](OH)SO_4$, on metathesis with barium *d*-tartrate and digestion in a steam bath, produces violet crystals of $[Co(OH)(H_2O)(bigH)_2]$*d*-tart.$3H_2O$ ($[\alpha]_D$ = (+) 460°; $[M]_D^{22^\circ C}$ = (+)2281°) [298]. Heating *cis*-$[Co(NH_3)_2(bigH)_2]Cl_3$ with sodiumnitrate yields ammonia gas and cis-$[Co(NO_2)_2(bigH)_2]^{3+}$ [299, 331].

Cis and *trans* isomers of $[CoX_2(bigH)_2]^{3+}$ have been synthesized and characterized (X = py, Me–NH_2 or γ-picoline). $[Co(Me\text{–}NH_2)_2(bigH)_2]Cl_3$ has been resolved by diastereoisomer formation with silver-*d*-tartrate and fractional crystallization of the chloride-*d*-tartrate. $[\alpha]_D^{18.5^\circ C}$ = (+)437.5° for the *d*-compound, indicating an asymmetric transformation of the second order [299, 331-333]. The transformation from the *l*- to the *d*-isomer occurs very quickly, and only one form (*d*) was isolated. Resolution of the bis(methylamine) complex into optical isomers indicates its *cis* configuration [299, 331].

Trans-$[Co(Me\text{–}NH_2)_2(bigH)_2]X_3$ or $[Co(py)_2(bigH)_2]X_3$ (X = Cl or Br) loses two molecules of methylamine or pyridine to yield uniformly bivalent cobalt complexes without destruction of the bond of the biguanide with the cobalt ion upon heating [333, 334]. Similar reactions have been observed on heating *trans*-$[Co(Me\text{–}NH_2)_2(Ph\text{–}bigH)_2]X_3$ or *trans*-$[Co(py)_2(Ph\text{-}bigH)_2]X_3$ (Fig. 85).

$$\text{rans-}[Co^{III}(Me\text{-}NH_2)_2(Ph\text{-}bigH)_2]Cl_3\cdot 2H_2O \xrightarrow[-2Me-NH_2]{-2H_2O} [Co^{II}Cl(Ph\text{-}big.HCl)_2]$$

$$\text{trans-}[Co(py)_2(Ph\text{-}bigH)_2]X_3\cdot 2H_2O \xrightarrow[-2\,py]{-2H_2O} [Co^{II}X(Ph\text{-}big\cdot HX)_2]$$

Fig. 85

The reduction of cobalt(III) to cobalt(II) can be explained as due to electron transfer from the two π-electron rich Ph-bigH to the central cobalt ion, a view supported by IR and far-IR data. The bands at 1,572 and 1,267 cm^{-1} in $[CoCl(Ph–bigH.Cl)_2]$, assigned to the stretching vibrations of C = N and C–N bonds, respectively, shift to lower wavenumbers by about 20 cm^{-1}. The shift may arise from the decrease of π-electron delocalization in the Ph-bigH, due to the electron transfer from Ph-bigH to the cobalt(III) ion. The decrease in the stretching vibration of Co–N (Ph-bigH) is also observed in the far-IR region upon heating (510 sh ⟶ 502 cm^{-1}). The decrease may result from the difference in the force constants of Co(III)–N and Co(II)-N bonds. The electronic spectrum of solid *trans*-$[Co(Me–NH_2)_2(Ph–bigH)_2]Cl_3.2H_2O$ changes remarkably on heating. The distinct feature of the spectra commonly found in these complexes before heating is the strong charge transfer band which obscures the second absorption band, suggesting strong π-bond character of the Ph-bigH-metal bond [334].

Sodium azide reacts with *cis*-$[Co(NH_3)_2(bigH_2]Cl_3$ to yield red-violet crystals of *cis*-$[Co(N_3)_2(bigH)_2]N_3.H_2O$. The corresponding *trans* isomer was prepared by addition of excess sodium azide to *trans*-$[CoCl_2(bigH)_2]Cl$. The *cis* isomer shows two peaks at 296 nm (33, 783 cm^{-1}; ε = 316) and 508 nm (19,685 cm^{-1}; ε = 168), while the *trans* isomer shows two peaks at 350 nm (28,571 cm^{-1}; ε = 208) and 482 nm (20,746 cm^{-1}; ε = 168). The bending mode $\delta(N_3)$ and the asymmetric stretch of coordinated azides υasy (N_3^-) were detected, but the assignment of the symmetrical band νasy (N_3^-) is doubtful as claimed by the authors [300].

A mixed complex, $NO[Co(NO)_2(big)_2].3H_2O$, separated when nitric oxide gas was passed through an aqueous suspension of $[Co(bigH)_2]^{2+}$. This diamagnetic compound contains NO^+ and NO^- radicals which are *cis*-disposed (XXXVII) (Fig. 86).

Air oxidation of bis(biguanide)cobalt(II) in the presence of β-alanOH deposited $[Co(OH)(\beta\text{-alan-}\bar{O})]^+$. In this compound β-alanOH is linked to the cobalt(III) ion by the N atom only (monodentate ligand). The action of potassium cyanide on the $[Co(bigH)_3]^{3+}$ ion forms $[Co(CN)_2(big)(bigH)]$. A similar

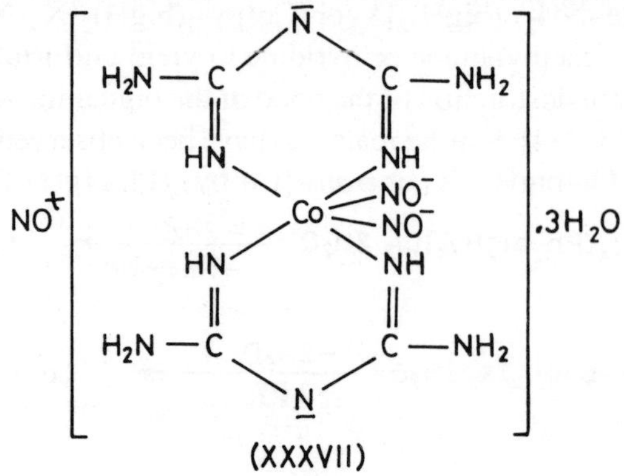

(XXXVII)
Fig. 86

compound $[Co(CN)_2(Pr\text{-}big)(Pr\text{-}bigH)].H_2O$ has also been reported [194]. Compounds of the types $[Co(L)_2\,en(bigH)_2]^{3+}$ (L = NH_3, Me–NH_2, Et NH_2 or benzylamine) and $[Co(py)(OH)\,en(bigH)_2]SO_4.3H_2O$ have been prepared and characterized [100].

Rây and Saha isolated $[Co(NO)(H_2O)\,en(bigH)_2]Br_2$ and $[Co(NO)(H_2O)\,en(bigH)_2]SO_4.H_2O$. The first compound shows five bands at 17,391 cm^{-1} (ε = 211); 20,080 cm^{-1} (ε = 213); 28,238 cm^{-1} (ε = 235); 37,000 cm^{-1} (sh, ε = 379); and 44,444 cm^{-1} (ε = 411), which are identified as ${}^1A_{1g} \longrightarrow {}^1T_{1g}$, ${}^1A_{1g} \longrightarrow {}^1T_{2g}$ charge transfer bands, and $\pi \rightarrow \pi^*$ transition. Thermal analysis of $[Co(NO)(H_2O)\,en(bigH)_2]\,SO_4.H_2O$ shows that at ~ 100°C the lattice water is lost and at ~ 180°C both nitric oxide and water molecules are removed. At 250°C the compound totally decomposes to a polymeric material [305].

The acid dissociation constant of cobalt(III) mixed chelates such as $[Co(gly\text{–}\bar{O})(bigH)_2]^{2+}$ (pK_1 = 9.06 ± 0.06; pK_2 = 10.36 ± 0.08), $[Co(alan\text{–}\bar{O})(bigH)_2]^{2+}$ (pK_1 = 9.07 ± 0.05, pK_2 = 10.41 ± 0.04), $[Co\,(val\text{–}\bar{O})\,(bigH)_2]^{2+}$ (pK_1 = 9.04 ± 0.02, pK_2 = 10.37 ± 0.03), and $[Co\,(sar)\,(bigH)_2]^{2+}$ (pK_1 = 8.84 ± 0.04, pK_2 = 10.32 ± 0.05) have been reported [65]. The acid dissociation constant of $[Co(H_2O)_2(bigH)_2]^{3+}$ at various temperatures were reported by Chakraborty [317] (Fig. 87).

pK_1 = 5.86 ± 0.90; pK_2 = 7.92 ± 0.05 at 35°C
pK_1 = 5.90 ± 0.07; pK_2 = 8.00 ± 0.04 at 40°C
pK_1 = 6.00 ± 0.13; pK_2 = 8.20 ± 0.02 at 50°C

Fig. 87

Absorption spectra and C.D. data for the $[Co\,(sar)\,(bigH)_2]^{2+}$ ion are also available [65].

(f) Cobalt(III) Heterochelates Containing One Bidentate Biguanide and Monodentate or Bidentate Ligands

Chakraborty [313] isolated and characterized several *cis*-(α-) and *cis*-(β) complexes of cobalt(III) with the unsymmetrical glycinato ligand, viz., violet and red $[Co(bigH)(gly\text{-}\bar{O})(OH)(H_2O)]ClO_4$ and violet and reddish-pink cis-(β)-dichloroglycinato(biguanide)cobalt(III), $[Co(bigH)(gly\text{–}\bar{O})Cl_2]$. The electronic spectra of the hydroxoaqua complexes show two bands at 18,900 cm^{-1} (${}^1A_{1g} \longrightarrow {}^1T_{1g}$) and 27,700 cm^{-1} (${}^1A_{1g} \longrightarrow {}^1T_{2g}$) for the violet compound and 20,000 cm^{-1} and 27,500 cm^{-1} for the reddish, pink compound, respectively (XXXVIII) (Fig. 88).

As the α-form is less symmetrical, it shows more IR bands than the corresponding β-form.

IR (α-form) : 3,460, 1,780, 1,760, 1,595, 1,565, 1,420, 1,310, 1,293, 625, 620, 520, 490, 432 cm^{-1}.

IR (β-form) : 3,500–3,180 (br), 1,700, 1,600 (br), 1,580, 1,530, 1,390, 1,250, 730, 620, 510 cm^{-1}.

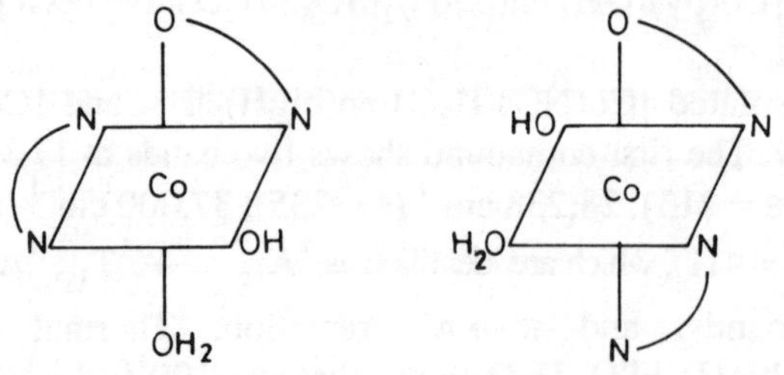

Fig. 88 (XXXVIII)

The complexes $[CoCl_2(LL)(paludH)_2]Cl$ (LL = *o*-phen or dipy, en) have been reported and characterized [335].

The complexes $[Co(bigH)(trien)]I_3$, $[Co(en)(pn)(bigH)]I_3.H_2O$, $[Co(en)(trien)(bigH)]I_3$ are novel because the metal atom is linked by three different bidentate ligands in the same coordination sphere [303, –305, 336]. The complexes $[CoCl_2(AA)(paludH)]Cl$ (AA = *o*-phen or dipy, en) were also reported and characterized [184].

(viii) Chromium(III)

Chromium(III)-biguanide/substituted biguanide complexes reported by Rây and others [1, 14] undergo hydrolysis, first forming $[Cr(OH)(H_2O)(bigH)_2]^{2+}$, then $[Cr(OH)_2(H_2O)_2(bigH)]^+$, and finally $[Cr(OH)_3(H_2O)_3]$ [1, 175, 191]. On heating at 150-160°C, the complex base $[Cr(big)_3].H_2O$ loses water to form the anhydro base, and on titration gives a sharp inflection at three equivalents of perchloric acid [1, 14, 100, 202, 215]. The stabilities of some substituted chromium(III)-biguanide complexes were reported by Poddar *et al* [215]. The ultraviolet and visible absorption spectra of chromium(III)-biguanide compounds have been reported [56, 78, 138, 147, 149, 153, 175, 337]. The authors [149] concluded that the absorption spectrum of $[Cr(big)_3]$ in the visible region is made up of three bands, of which the two longer wavelength bands are polarized at right angles to the shorter wavelength one. The large overall splitting of the T_{2g} levels in $[Cr(big)_3]$ may thus be due to a joint effect of π-bonding and Jahn-Teller distortion.

In aqueous solution $[Cr(bigH)_3]Cl_3$ shows two absorption bands at 482 nm (20,747 cm^{-1}) and at 369 nm (27,100 cm^{-1}). These transitions may be ascribed to $^4A_{2g} \longrightarrow {}^4T_{2g}$ and $^4A_{2g} \longrightarrow {}^4T_{1g}$ transitions. The anhydrous base $[CrL_3]$ shows two bands at 20,000 cm^{-1} and 26,000 cm^{-1}, whereas the corresponding hydrated base $[CrL_3].nH_2O$ shows bands at 19,700 cm^{-1} and 26,000 cm^{-1}. Small but definite band shifts to lower wavenumbers for the spin-allowed transitions occur on going from charged $[Cr(LH)_3]^{3+}$ to uncharged $[CrL_3]$. Such band shifting may be explained in terms of the relationship between the charge of the metal ion and the Δ value [147].

Michelsen [56] investigated the C.D. spectra of $(-)_{589}$ – $[Co(bigH)_3]^{3+}$ and $(-)_{589}$ – $[Cr(bigH)_3]^{3+}$ and assigned the same absolute configuration, Λ, to them. Brubaker and Webb [57] confirmed the absolute configuration of $(-)_{589}$ – $[Cr(bigH)_3]^{3+}$ to be Λ by X-ray analysis. From the comparison of the UV–C.D spectra of the cobalt(III) and chromium(III) complexes, they suggested that $(-)_{589}$ – $[Co(bigH)_3]^{3+}$ should have the Δ configuration. Under these circumstances, Igi *et al* [58] related the C.D. data for $(-)_{589}$ – $[Co(bigH)_3]^{3+}$ to those for $(+)_{589}$ – $[Co(en)_3]^{3+}$ by studying the mixed chelates$(+)_{589}$–$[Co(en)(bigH)_2]^{3+}$ and $(+)_{589}$ – $[Co(en)_2(bigH)]^{+3}$ and assigned the Λ absolute configuration to $(-)_{589}$ – $[Co(bigH)_3]^{3+}$. Several years later, this assignment was confirmed by Snow [30] using an X-ray technique. Kawaguchi and Douglas [62] studied absorption and C.D. spectra of [Co (*o*-phen or dipy) $(bigH)_2]^{3+}$ in water, 0.1 *M* sodium hydroxide, and ethanol. The hydroxyl ion affects the C.D. intensity mainly in the first absorption band. They also observed a marked solvent effect of ethanol for both absorption and C.D. spectra, especially in the first absorption band region. The absolute configurations for these complexes are assigned on the basis of their C.D. spectra. Further, $[Cr(bigH)_3]^{3+}$ has been used as a paramagnetic probe to perturb selectivity and assign the NMR spectra of iron(III)-protoporphyrin-IX-dicyanide (hemine-CN) [338].

The stepwise dissociation of $[M(bigH)_3]^{3+}$ (M = Co(III) or Cr(III)) was studied spectrophotometrically by Banerjea and Chakraborty [281].

In aqueous media, $[Cr(bigH)_3]^{3+}$ is susceptible to hydrolysis and reacts smoothly with picolinic/quinaldenic acid to form mixed chelates of the type $[Cr(AA)(bigH)_2]^{2+}$ (HAA = picolinic acid/quinaldenic acid) [191]. The compounds are paramagnetic, and their 10 Dq values are much greater than the values for $[Cr(H_2O)_6]^{3+}$ (17,400 cm^{-1}) and $[CrCl_2(H_2O)_4]^+$ (15,748 cm^{-1}), but comparatively less than that for $[Cr(en)_3]^{3+}$ (21,930 cm^{-1}) [191].

A kinetic and mechanistic study of the formation of $[Cr(AA)(bigH)_2]^{3+}$ (AA = *o*-phen or dipy) by reaction of *cis*-$[Cr(H_2O)_2(bigH)_2]^{3+}$ with *o*-phen or dipy was reported by Banerjea and Sengupta [339], who also prepared $[Cr(Ox)(bigH)_2]^+$ by reaction of $[Cr(H_2O)_2(bigH_2]^{3+}$ with sodium oxalate at pH~4 [339]. The much faster rate for *cis*-$[Cr(H_2O)_2(bigH)_2]^{3+}$ is presumably due to a much lower $\Delta H^{\ddagger}$ value (84.4 KJ $mole^{-1}$) compared to that for $[Cr(H_2O)_6]^{3+}$ ($\Delta H^{\ddagger}$ = 111.2 kJ $mole^{-1}$) for the biguanide complex compared to that of $[Cr(H_2O)_6]^{3+}$ (30.9 KJ^{-1} $mole^{-1}$) [340]. The much lower $\Delta H^{\ddagger}$ value for the biguanide complex indicates considerable weakening of the Cr-OH_2 bond by the strong donor bigH.

Reaction of $[Cr(H_2O)_2(bigH)_2]^{3+}$ with *o*-phen or dipy also proceeds much faster [319] than that of $[Cr(H_2O)_6]^{3+}$ [321]. For the reaction of *cis*-$[Cr(H_2O)_2(bigH)_2]^{3+}$ forming the mixed chelates $[Cr(AA)(bigH)_2]^{n+}$ (AA = dipy, *o*-phen, or Ox) the $\Delta H^{\ddagger}$ values are nearly identical (84.8 ± 1.5 KJ $mole^{-1}$), indicating dissociative processes; the first-order dependence of rate on ligand concentration in each case is due to the formation of an outer-sphere precursor. The dissociative activation (Id) in these cases is possibly due to the presence of the strong donor biguanide, which causes much labilization of the Cr–OH_2 bonds, thus favouring a dissociative mechanism and a much faster substitution than in the case of $[Cr(H_2O)_6]^{3+}$. In fact, the lability of the aqua ligands for substitution by dipy or *o*-phen is much greater in *cis*-$[Cr(H_2O)_2(bigH)_2]^{3+}$ than in even the anionic *cis*-$[Cr(H_2O)_2(Ox)_2]^-$ [321]. Acid hydrolysis of $[Cr(Ox)(bigH)_2]^+$ shows that the two bigH molecules are lost in two successive stages. At low acid concentrations (0.01-0.06 *M*) $[Cr(H_2O)_2(Ox)(bigH)]^+$ is formed, which may further dissociate at much higher acid concentrations to form $[Cr(Ox)(H_2O)_4]^+$ [339].

Rây and Saha [341] prepared dithiocyanatobis(biguanide)chromium(III) thiocyanate and formulated the compound as $[Cr(NCS)(SCN)(bigH)_2]SCN$ without any physical data. A modified preparative technique for its synthesis was described by Syamal [342]. The compound is paramagnetic and shows two absorption bands at 20,410 cm^{-1} (10 Dq value) and at 26,660 cm^{-1}. The IR spectra indicate that the two coordinated thiocyanate ions are bonded to chromium(III) through the N atoms and not through the S atoms. The Cr-N stretch has also been detected at 350 cm^{-1}.

Dutta and Ray [150, 343] prepared and characterized the following compounds: $[Co(o\text{-phen})(bigH)_2](OH)Cl_2$, $[Cr(o\text{-phen})(bigH)_2](OH)SO_4.2.5H_2O$, $[Cr(o\text{-phen})(bigH)_2]Cl_3.3H_2O$, $[Cr(o\text{-phen})(bigH)_2]Cl(SO_4)$, $[Cr(o\text{-phen})(bigH)_2](NO_3)SO_4.4H_2O$, $[Cr(o\text{-phen})(bigH)_2](SO_4)_{1.5}.5H_2O$, $[Co(dipy)(bigH)_2](OH)(BPh_4)_2$, $[Co(PBz)_2(bigH)]Cl$, $[Cr(dipy)(bigH)_2](SO_4)_{1.5}.1.5H_2O$ and [Cr(bigH)(dipic)(dipic ... H)].

In aqueous media, [Cr(bigH)(dipic)(dipic-H)] dissociates, as confirmed by conductance measurements (64 ohm^{-1} cm^2 $mole^{-1}$). An oxygen-dominated chromophore, *viz.*, $[CrN_4O_2]$ is revealed by its bluish-purple colour and its absorption maximum at 18,200 cm^{-1}, substantially lower than the absorption of $[Cr(pic)(bigH)_2]Cl_2$ (20,800 cm^{-1}) with a $[CrN_5O]$ chromophore. $[Cr(PBz)_2(bigH)]Cl$ (PBz = 2-(2′-pyridylbenzimidazole) shows a band at 18,200 cm^{-1}, which is lower than those observed from several other $[CrN_6]$ complexes. Deprotonation of PBzH thus appears to be a weaker ligand compared to bigH, en, or NH_3 (Table 18) [150] :

Compounds of type $K[Cr(NCS)_4(R\text{-bigH})]$ (R-bigH = Ph-bigH, *p*-Cl-Ph-bigH, *o*-tolyl-bigH, or paludH) were prepared by the action of the substituted biguanide on $K_3[Cr(NCS)_6]$ in alcoholic medium at *p*H ~ 7. The Cr-N bonding

is confirmed by IR measurements [346]. The syntheses and characterization of the following compounds were reported [344, 346-349] :

Table 18 Electronic Spectra of Some Chromium (III) Mixed Complexes

Complex	*State*	λ_{max} *(nm)*	ε	*Ref.*
$[Cr(NCS)_2(bigH)_2]NCS$	mull	390	–	342
		375	–	
$[Cr(dipy)(bigH)_2](SO_4)_{1.5}$	water	470		343
$[Cr(dipy)(bigH)_2]ClO_4)_3$	water	470		343
[Cr(*o*-phen)$(bigH)_2](ClO_4)_3$	water	470		343
$[Cr(pic)(bigH)_2]Cl_2$	water	480	84	191
[Cr(bigH)(dipic)(dipic ... H)]	water	549		343
$[Cr(quinal)(bigH)_2](ClO_4)_2$	water	492	93	191
$[Cr(H_2O)_2(bigH)_2]^{3+}$	water	{490, 385		339
$[Cr(H_2O)_4(bigH)]^{3+}$	water	490		339
$[Cr(Ox)(bigH)_2]^+$	water	{490, 380		339
$[Cr(Ox)(H_2O)_2(bigH)]^+$	water	490		339
$[Cr(PBz)_2(bigH)]Cl$	water	490		150
$[Cr(IDA)(bigH)_2]ClO_4$	water	490		344, 345
$[Cr(histOH)(bigH)_2](ClO_4)_3$	water	490		320

$K_2[Cr(NCS)_5$(*o*-tolyl-bigH)]$.2H_2O$, K$[Cr(NCS)_4$(*o*-tolyl-bigH)], $[Cr(NCS)_2$-(*o*-tolyl-big)(*o*-tolyl-bigH)], (*o*-tolyl–$bigH_2)[Cr(NH_3)_2(NCS)_4]$, (diphenyl-$bigH_2)[Cr(NH_3)_2(NCS)_4]$, (diphenylbig$H_2)_3[Cr(NCS)_6]$ and [Cr(*o*-tolyl-$bigH)_3](NO_3)_3$.

The dissociation of $[Cr(R\text{-}bigH)_3]^{3+}$ in acid ($HClO_4$) medium was first order with respect to both the complex and the hydrogen ion concentrations (R-bigH = bigH, Ph-bigH, or *n*-hex-bigH) [281].

Protonation might be expected to occur at the free basic =NH group of the metal-bonded biguanide (XXXIX) (Fig. 89) :

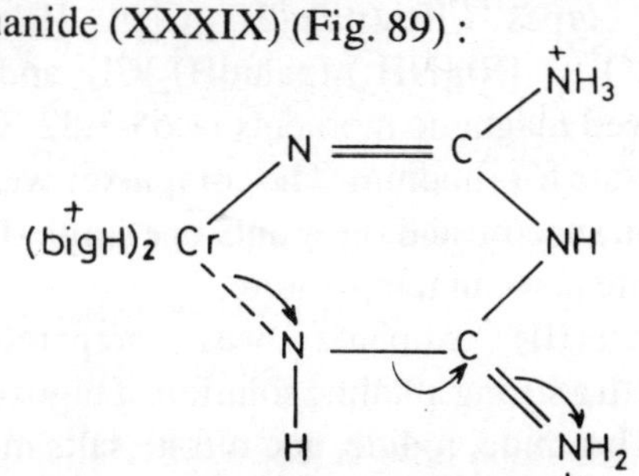

Fig. 89 (XXXIX)

Protonation of the metal-bonded ligand will facilitate bond breaking by exerting a pull on the electrons of the outer shell of the adjacent atom, and the effect will be relayed through the molecule to weaken the metal-ligand bond, which will break. The rate of dissociation of $[Cr(bigH)_3]^{3+}$ in alkaline media decreases with increase in the concentration of hydroxide ion, and a protonation mechanism is suggested [281] (Fig. 90).

$$[Cr(bigH)_3]^{3+} + 3OH^- \longrightarrow [Cr(big)_3] + 3H_2O$$

$$[Cr(big)_3] + H_2O \xrightarrow{\text{fast}} [Cr(bigH)(big)_2]^+ + OH^-$$

$$[Cr(bigH)(big)_2]^+ \xrightarrow[+\,H_2O]{\text{slow}} [Cr(H_2O)_2(big)_2]^+ + bigH$$

Fig. 90

The rate of formation of the chromium(III)-EDTA complex in the reaction of $[Cr(bigH)_3]^{3+}$ with EDTA in alkaline media (*p*H 7.4 to 9) has been followed spectrophotometrically [327]. The reaction followed two concurrent paths—one independent of *p*H and $EDTAH_4$ concentration and the other showing an inverse first-order dependence of H^+ activity and first-order dependence on $[EDTAH_4]$. The metal ion-catalyzed dissociation of $[Cr(bigH)_3]^{3+}$ into $[Cr(H_2O)_2(bigH)_2]^{3+}$ in acidic media proceeds through the formation of an intermediate, in which the catalyzing metal ion is chelated to the coordinated biguanide of the substrate complex. The catalytic activity of the metal ion decreases in the order:

$$Cu^{2+} > Ni^{2+} > Co^{2+} > Zn^{2+} > Mn^{2+}$$

The kinetics of substitution of aqua ligands from *cis*-$[Cr(H_2O)_2(bigH)_2]^{3+}$ by picH, *l*-histOH, $AspH_2$, and $IDAH_2$ in aqueous or $EtOH–H_2O$ media has been reported [320, 344, 345]. The reaction rate is *p*H-dependent. The results suggest that the reaction occurs by an associative interchange process via ion-pair formation. In this process, both bond breaking and bond making are equally important. The energy of bond breaking is thus partially compensated for by the energy released by the bond formation by the incoming ligand in the transition state. This is supported by a low $\Delta H^{\ddagger}$ value (20.4 kcal/mole).

(ix) Rhodium(IV) and (III)

Complexes of the types $[Rh(paludH)_3]Cl_4$, $[RhCl_2(paludH)_2]Cl_2$, $[Rh(NO_2)_2(paludH)_2](NO_2)_2$, $[Rh(NH_3)_2(paludH)_2]Cl_4$ and $[Rh(py)_2(paludH)_2]Cl_4$ [258, 350] showed magnetic moments (1.65-1.82 B.M.) supporting a quadrivalent oxidation state for rhodium. The complexes were chracterized by ultraviolet and absorption spectra, and show only one single-line EPR spectrum in the polycrystalline state at room temperature.

Tris(biguanide)rhodium(III) sulphate was prepared by refluxing rhodium(III) chloride with a strong alkaline solution of biguanide sulphate. The corresponding chloride, bromide, iodide, and nitrate salts may be synthesized by metathesis. TG studies of $[Rh(bigH)_3]_2(SO_4)_3.3.5H_2O$ indicate that at approximately 150°C all the water of crystallization is lost, yielding $[Rh(bigH)_3](SO_4)_{1.5}$. At 280°C the anhydrous salt starts to decompose, and at 650°C the compound is converted to rhodium(III) oxide. $[Co(bigH)_3](SO_4)_{1.5}$ decomposes at 260°C indicating that the rhodium(III) compound is more stable than the corresponding cobalt(III) compound [351]. The compounds are diamagnetic and show one absorption band at 308 nm (32,470 cm^{-1}), $^1A_{1g} \longrightarrow {}^1T_{1g}$ transition [154, 155].

$[Rh(bigH)_3]Cl_3$ was resolved by forming the diastereoisomer, (+)-tris-(biguanide)rhodium(III) chloride (+)–tartrate, which was then fractionally crystallized. The least soluble fractions are *levo*-rotatory. Unlike the diastereoisomer $[Co(bigH)_3]Cl$ (+)–tart, this rhodium analogue does not show any "asymmetric transformation of the second order." The *levo*-forms are the less soluble variety and separate first. The molecular rotation of $[Rh(bigH)_3]Cl$-(+)–tart, $[\alpha]_D = (\pm)140°$, $[M]_D^{32°C} = (\pm)951°$) is much less than that of the corresponding cobalt(III) complex ($[M]_D^{32°C} = (\pm)\ 2153°$). The rotations of related salts are (Fig. 91) :

$$l\text{-}[Rh(bigH)_3]Cl_3,\ [\alpha]_D = (-)155°,\ [M]_D^{32°C} = (-)794°$$
$$d\text{-}[Rh(bigH)_3]Cl_3,\ [\alpha]_D = (+)150°,\ [M]_D^{32°C} = (+)768.5°$$
$$l\text{-}[Rh(bigH)_3]Cl_3,\ [\alpha]_D = (-)135°,\ [M]_D^{32°C} = (-)799°$$

Fig. 91

Optically active $[Rh(bigH)_3]X_3$ salts show no tendency to racemize even on standing for years in the solid state or in aqueous solution. Different anions and cations have no effect on their inertness, and racemization could not be effected even by boiling an aqueous solution of the chloride with activated charcoal. The higher molecular rotation of $[Co(bigH)_3]^{3+}$ compared to $[Rh(bigH)_3]^{3+}$ is due to the greater asymmetry of the $[Rh(bigH)_3]^{3+}$ cation. Like ethylenediamine, biguanide is symmetrical but coordinates unsymmetrically to the central metal ion [352]. Using substituted biguanides (R_2bigH or R-bigH = Me_2-bigH, morph-bigH or. palud), diamagnetic compounds of the types $[Rh(R\text{-}bigH)_3]X_3$, $[Rh(R\text{-}bigH)_2X_2]X$, $[Rh(paludH)_2Cl_2]$ $[Rh(paludH)Cl_4]$, $[Rh(NO_2)_2(paludH)_2]$ NO_2, $(paludH_2)[RhCl_4(paludH)]$, $(paludH_2)_2[RhCl_5(H_2O)]$ and $(R\text{–}bigH_2)_3$ $[RhCl_6]$ have been prepared [155, 353, 354]. $[Rh\ (paludH)_3]\ Cl_3$ has been resolved into *d*- and *l*-forms using silver tartrate as the resolving agent.

The tris (chelate ligand) rhodium (III) complexes show two bands—one at ~ 22,990 cm^{-1} ($^1A_{1g} \longrightarrow {}^1T_{1g}$) and the other at 31,250 cm^{-1} ($^1A_{1g} \longrightarrow {}^1T_{2g}$) respectively [155]. Rhodium complexes with Ph-bigH, *p*-Cl-Ph-bigH and tolyl-bigH [353], and *cis* and *trans* isomers of $[RhCl_2(paludH)_2]Cl$ were also reported [155]. The synthesis and characterization of $[RhX_2(bigH)_2]^+$ (X = Cl, Br, I, NCS, N_3, or NO_2) have been described by Ghosh and Bhattacharjee [299, 331, 355]. The compounds are diamagnetic and octahedral, and they have a *cis* configuration. Depending upon the *p*H of the solution, mono, bis, and tris compounds are formed.

When rhodium(III)chloride is refluxed with bigH.HCl in a 1 : 3 mole ratio in a highly acidic medium (8-10 *N* hydrochloric acid), a carmine-red compound, (bigH)H_2[$RhCl_4$(bigH)]Cl, is precipitated. If the *p*H of the solution is 2.5, a different, cream-yellow compound, [$RhCl_2$$(bigH)_2$]Cl.2$H_2O$, results. On further increase in *p*H to 3.0, the solution becomes colloidal, and at *p*H 7, the cream-white [Rh$(bigH)_3$]Cl_3 separates. The change from the monobiguanidinium complex to the bis(biguanidium) complex is reversible (Fig. 92).

$$\underset{\text{(carmine-red)}}{(\text{bigH})\text{H}_2[\text{RhCl}_4(\text{bigH})]_2} \underset{\text{warm with conc.HCl}}{\overset{\text{pH} \sim 2.5}{\rightleftharpoons}} \underset{\text{(cream-yellow)}}{[\text{RhCl}_2(\text{bigH})_2]\text{Cl}}$$

Fig. 92

However, once the tris compound is formed, it has no tendency to change to the bis compound when the *p*H is lowered. The stability of the complexes follows the order: $[RhCl_4(bigH)]^-$ < $[RhCl_2(bigH)_2]^+$ < $[Rh(bigH)_3]^{3+}$. The $[RhCl_2(bigH)_2]^+$ cation undergoes aquation as shown by the shifting of the absorption maximum from 332 nm (30, 120 cm^{-1}) to 316 nm (31,645 cm^{-1}), and the change in conductivity from 297 to 428 ohm^{-1} cm^2 $mole^{-1}$, which supports the reaction (Fig. 93) [355] :

$$[\text{RhCl}_2(\text{bigH})_2]\text{Cl} + 2\text{H}_2\text{O} \rightleftharpoons [\text{Rh}(\text{bigH})_2(\text{H}_2\text{O})_2]\text{Cl}_3$$

Fig. 93

In *cis*-[Rh$(NCS)_2$$(bigH)_2$]SCN the ν (C–N) stretching frequencies at 2,110 cm^{-1} and 2,042 cm^{-1} and the ν (C–S) stretching frequencies at 810 cm^{-1} and 776 cm^{-1} support Rh–NCS bonding [355, 356]. The IR spectrum of the $[Rh(N_3)_2(bigH)_2]^+$ cation shows a ν asy stretch at 2,038 cm^{-1}, whereas the ν asy stretches for [Rh$(N_3)_2$$(en)_2$]Cl, NH_4N_3, and CH_3N_3 are at 2,060 cm^{-1}, 2,030 cm^{-1} and 2,143 cm^{-1}, respectively. Therefore, the M–N bond in [Rh$(N_3)_2$$(bigH)_2$]Cl is more ionic than that in [Rh$(N_3)_2$$(en)_2$]Cl. The *cis* structure of [Rh$(NO_2)_2$$(bigH)_2$]$NO_3$ is confirmed by IR data. The sharp band at 1,350 cm^{-1}, and the two sharp ones at 835 cm^{-1} and 829 cm^{-1} also indicate a *cis* structure. The spectrum is almost identical with that of *cis*-[Co$(NO_2)_2$(bigH)]NO_3. The *cis* structure of $[RhX_2(bigH)_2]^+$ is substantiated by the substitution of two *cis*-chloro ligands in [$RhCl_2$$(bigH)_2$]Cl by picH, *o*-phen, or bigH. Attempts to resolve it into optical isomers by diastereoisomer formation were unsuccessful, and no corresponding *trans* isomer could be isolated [355]. The synthesis and characterization of [Rh_2{en$(bigH)_2$}$_3$]$(SO_4)_3$, *trans*-[RhX_2 en$(bigH)_2$]X, [RhX_2 pip$(bigH)_2$] X (X = Cl, Br, I, NO_2, NCS, or N_3) have been reported [357].

$[Rh(bigH)_3]^{3+}$, unlike $[Co(bigH)_3]^{3+}$, does not react with potassium cyanide, a fact explained by the larger ligand field splitting energy and higher stability of the rhodium complex. However, the action of potassium cyanide on $[RhCl_2(bigH)_2]Cl$ in basic medium favours the formation of $[Rh(CN)_2(big)(bigH)].H_2O$ or $[Rh(CN)_2(bigH)_2]OH$. Prolonged digestion of $[Rh(CN)_2(big)(bigH)]$ with 2 *N* hydrochloric acid favours the formation of $[Rh(CN)(AUH)_2]_2Cl_4.5H_2O$. No changes in electronic spectra (solid) were observed in the visible region (350-600 nm), but in the ultraviolet the complexes display a strong absorption below 340 nm (29,411 cm^{-1}) because of back bonding from metal to the anti-bonding orbital of the CN group. The complexes are diamagnetic and stable in air [343].

The bonding of the CN group through the carbon atom in $[Rh(CN)_2(big)(bigH)].H_2O$ is supported by a strong and sharp band at 2,100 cm^{-1}. Though ionic, $[Rh(CN)(AUH)_2]_2Cl_4.5H_2O$ is sparingly soluble in water, suggesting a dimeric structure. The carbon-bonded, bridging CN group is supported by IR spectra; a strong band at 2,175 cm^{-1} is observed. The ν(CO) stretch of AUH occurs at 1,680 cm^{-1}, and ν (NCO) vibrations occur at 1,070 cm^{-1}. A sharp peak at 485 cm^{-1} and a medium band at 530 cm^{-1} are assigned as ν (Rh-C) and ν(Rh–N) stretches. In the case of $[Rh(CN)(AUH)_2]_2Cl_4.5H_2O$, a medium band at 570 cm^{-1} and a weak band at 450 cm^{-1} were tentatively assigned to ν(Rh–N) and ν(Rh-C), respectively [358].

The acid perchloric acid-catalyzed dissociation of $[Rh(bigH)_3]^{3+}$ into $[Rh(H_2O)_2(bigH)_2]^{3+}$ has been studied [282]. The observed dependence of the rate on the acid concentration and the Hammett acidity function (H_o) suggest a SN_1 CA mechanism. The plot of log R_{obs}. vs. $(- H_0)$ is linear with a slope ≈ 1 in the range of 0.8-3.0 *M* perchloric acid, with $\omega = - 3.5$ and $\phi = -0.43$. The activation parameters of this reaction show that $\Delta H^{\ddagger}$ values are lower for the reactions of cobalt(III) and tris(biguanide)chromium(III) complexes because of the lower strength of the metal-ligand bond compared to the rhodium(III)-biguanide complexes [282].

The aquation of *cis*-$[RhX_2(bigH)_2]NO_3$ (X = Cl, Br, or I) was investigated conductimetrically [359]. The amount of X^- released during the reaction, measured as AgX, indicated that the reaction proceeds almost completely in the forward direction, forming $[Rh(H_2O)_2(bigH)_2]^{3+}$. It does not follow simple first-order kinetics, and on this basis it is assumed that the reaction occurs in a stepwise manner (Fig. 94) :

$$[RhX_2(bigH)_2]NO_3 + H_2O \xrightarrow{k_1} [RhX(H_2O)(bigH)_2]NO_3.X$$

$$[RhX(H_2O)(bigH)_2]NO_3.X + H_2O \xrightarrow{k_2} [Rh(OH)(H_2O)(bigH)_2]NO_3.X + HX$$

Fig. 94

For the first step of anation, metal-halide bond rupture is the only important factor and is virtually complete in the transition state. For the second step of anation, bond formation by the aqua ligand occurs before the bond rupture is complete. The loss of the second halide ion takes place along with a hydrogen ion from the "aqua" ligand with no resultant formation of the hydroxoaquabis (biguanide) complex [359] (Fig. 95) :

$$[RhX(H_2O)(bigH)_2]NO_3.X + H_2O \xrightarrow{k_2} [Rh(OH)(H_2O)(bigH)_2]NO_3.X + HX$$

Fig. 95

Studies on the aquation of $[RhCl_2(R\text{-}bigH)_2]NO_3$ (R-bigH = Me-bigH, Et-bigH, or Ph-bigH) were carried out conductometrically in aqueous and aqueous-methanol media [131]. In aqueous media, the rate of aquation is smaller for $[RhCl_2(R\text{-}bigH)_2]NO_3$ than for the corresponding unsubstituted complex $[RhCl_2(bigH)]NO_3$. However, in aqueous-methanol media, the rate decreases with increase in the percentage of methanol for the unsubstituted biguanide complex, but the reverse is true for Ph-bigH complex because of its greater solvation in the organic-enriched solvent.

(x) Iridium(IV) and (III)

Iridium(IV) compounds of the types $[Ir(paludH)_3]Cl_4$, $[Ir(NH_3)_2(paludH)_2]Cl_4$, $[Ir(py)_2(paludH)_2]Cl_4$, $[IrCl_2(paludH)_2]Cl_2$, $(Me_2\text{–}bigH_2)_2[IrCl_6]$, $(morph\text{–}bigH_2)_2[IrCl_6]$ and $(paludH_2)_2[IrCl_6]$ were reported by Spacu and Gheorghiu [136, 346]. The complexes were prepared by oxidation of iridium(III) chloride with sodium chlorite, followed by addition of alcoholic solutions of the ligands. Depending upon the *p*H of the solution, different compounds separated, e.g., at *p*H 7, $[Ir(paludH)_3]Cl_4$ is formed, while around *p*H 5 $(Me_2\text{–}bigH_2)_2[IrCl_6]$ resulted. The compounds are stable both in the solid state and in aqueous solution. Their magnetic moment values (1.56-1.80 B.M.) correspond to iridium(IV) and show a *d-d* band at 21,280-21,740 cm^{-1} [136]. On heating $(Me_2\text{–}bigH_2)_2[IrCl_6]$ or $(morph\text{–}bigH_2)_2[IrCl_6]$ at 120-140°C, thermal electron transfer occurs, and iridium (IV) is reduced to iridium(III) with the evolution chlorine and hydrochloric acid (Fig. 96) :

$$(R\text{-}bigH_2)_2[IrCl_6] \xrightarrow{\Delta} [IrCl_2(R\text{-}bigH)_2]Cl + 2HCl + 1/2Cl_2 \uparrow$$

Fig. 96

Spacu and Georghiu [136] measured the absorption spectra of many of these iridium(IV) complexes.

The preparation and properties of $[Ir(bigH)_3]^{3+}$ were first reported by Sinha and Ghosh [156-158]. The thermal stability of

$$\underset{(330°C)}{[Ir(bigH)_3]Cl_3} > \underset{(320°C)}{[Rh(bigH)_3]Cl_3} > \underset{(265°C)}{[Co(bigH)_3]Cl_3}$$

and the residue is presumed to be iridium (III) oxide.

$[Ir(bigH)_3]Cl_3$ has been resolved into its optical isomers by fractional crystallization of the diastereoisomers, $[Ir(bigH)_3]Cl.(+)$–tart, which does not exhibit "asymmetric transformation of the second order". The *levo-* compound is the less soluble variety and separates first. The (–) and (+) enantiomers of the complex chloro (+)-tartrate, (–) and (+) complex chloride, and nitrate, have been obtained in a pure state, and their optical rotations were measured (Fig. 97).

l-$[Ir(bigH)_3]Cl$ (+)-tart	$[\alpha]_D = (-)125°$, $[M]_D = (-)893°$
d-$[Ir(bigH)_3]Cl$ (+)-tart	$[\alpha]_D = (+)130°$, $[M]_D = (+)928°$
l-$[Ir(bigH)_3]Cl_3$	$[\alpha]_D = (-)120°$, $[M]_D = (-)722°$
d-$[Ir(bigH)_3]Cl_3$	$[\alpha]_D = (+)110°$, $[M]_D = (+)662°$
l-$[Ir(bigH)_3](NO_3)_3$	$[\alpha]_D = (-)100°$, $[M]_D = (-)681°$
d-$[Ir(bigH)_3](NO_3)_3$	$[\alpha]_D = (+)102°$, $[M]_D = (+)694°$

Fig. 97

Iridium(III) complexes have been reported with substituted biguanides (Me_2bigH, morph-bigH, and paludH) [159]. $[Ir(paludH)_3]^{3+}$ has been resolved into both optical isomers via the chloride *d*-tartrate diastereoisomers. For the *l*-and *d*-forms of $[Ir(bigH)_3](NO_3)_3$, $[\alpha]_D$ = (–)73°, $[M]_D$ = (–)831.5° and $[\alpha]_D$ = (+)75° and $[M]_D$ = (+)854°, respectively [136]. Gheorghiu [136] also prepared compounds of the types: $[Ir(R\text{–}bigH)_3]X_3$ (X = Cl, Br, I, CNO, N_3, NO_2, NO_3, $0.5SO_4$ or tart), $[Ir(R\text{–}bigH)_2Cl_2]Cl$, $[Ir(NH_3)_2(R\text{–}bigH)_2]Cl_3$, $[Ir(py)_2(R\text{-}bigH)_2]Cl_3$ (R-bigH = Me_2-bigH, morph–bigH, or paludH), $[IrCl_2(paludH)_2]$ $[IrCl_4(paludH)]$, $[IrCl_2(paludH)_2]Cl$, $[Ir(NH_3)_2(paludH)_2]Cl_3$, $[Ir(py)_2(paludH)_2]Cl_3$, $(paludH_2)[IrCl_4(paludH)]$, $(paludH_2)_3[IrCl_6]$ and $(paludH_2)_2[IrCl_5(H_2O)]$. The structures of the complexes were established by chemical methods and electronic spectra, and they were resolved into their optical antipodes.

Ghosh and Bhattacharjee [299] prepared *cis-* and *trans-*$[IrX_2(bigH)_2]X$ (X = Cl, Br, I, NCS, or NO_2). The *cis* compounds undergo substitution reactions with *o*-phen, or dipy with the formation of the heterochelates [Ir(*o*-phen)$(bigH)_2]Cl_3$ or $[Ir(dipy)(bigH)_2]$ Cl_3. $[Ir(CN)_2(big)(bigH)].H_2O$ has recently been isolated and characterized [358].

(xi) Ruthenium(III) and (II)

Ghosh and Prasad prepared a number of compounds with bigH and $en(bigH)_2$ [360]. $[RuCl_2(bigH)_2]X$ (X = Cl, Br or I) are paramagnetic and show only one broad band at 370 nm. $[Ru(CO)_2(bigH)_2]Cl_3$ shows a charge transfer band and a ν(CO) stretch at 1,960 cm^{-1}. $[Ru_2Cl_4\{en(bigH)_2\}(CO)_2]$ has a μ_{eff} of 1.69 B.M., suggesting Ru-Ru interaction. Its ν (CO) stretch is observed at ~ 1,900 cm^{-1}. The reflectance spectrum of $[Ru_2\{en(bigH)_2\}_2](SO_4)_3.5H_2O$

shows a shoulder at 440 nm, while that of $[Ru_2Cl_4\{en(bigH)_2\}(CO)_2]$ shows a sharp band at 400 nm (*d-d* transition) [360].

Fusing ruthenium(III) chloride with $bigH.H_2SO_4$ at 150-155°C produced a dirty yellow mass which, after extraction with water and ammonia, yielded yellow crystals of composition $[Ru(bigH)_3]_2(SO_4)_3.7H_2O$ [349]. The ompound is paramagnetic (μ_{eff} = 1.91 B.M.), as expected for d^5 low-spin ruthenium(III) complexes. The preparative method put forward by Sen [361] has been superseded by a completely different procedure by Ghosh and Prasad [360].

$[Ru(H_2O)_2(paludH)_2]Cl_3.10H_2O$, $[Ru(paludH)_3]X_3$ (X = Cl, N_3, OCN, SCN, 0.5 SO_4^{2-}, 0.5 *d*-tart, 1/3 $[Cr(NCS)_6]^{3-}$, or $[Cr(NH_3)_2(NCS)_4]^-$) have also been prepared. The compounds are soluble in water and methanol, are stable both in the solid state and in solution, and are photosensitive and paramagnetic [362].

Aquation of *cis*-$[Ru^{III}(bigH)_2Cl_2]Cl$ in aqueous-organic solvents has been recently reported [363]. The reaction occurs in steps with the formation of $[Ru(bigH)_2(H_2O)Cl]^{2+}$ and finally $[Ru(bigH)_2(H_2O)_2]^{3+}$. As the percentage of water molecules in the mixed solvent decreases, the rate of aquation also decreases. Activation parameters indicate a redox system in the first step, followed by an aquation process.

Complexes of the type $[Ru(paludH)_3]X_2$ (X = Cl, N_3, OCN, SCN, $0.5SO_4^{2-}$, 0.5 *d*-tart) are stable both in the solid state and in solution at *p*H 5-7.5. The +2 oxidation state for ruthenium is confirmed by the diamagnetism of the compounds [362].

(xii) Silver(III), (II), and (I)

Chakraborty and Rây [110, 111] prepared $[Ag^{III}en(bigH)_2]X_3$ (X = NO_3, ClO_4, OH, $0.5SO_4^{2-}$) by the oxidation of silver(I) salts by peroxydisulphate in the presence of the ligand at *p*H ≥ 6.5. The crystallization procedure reported earlier [111, 364] produces a complex with irreproducible spectral characteristics. Banerjee *et al* [365] modified the recrystallization procedure and recrystallized the nitrate by the addition of cold (excess) aerated 5% (v/v) nitric acid solution instead of 2*N* nitric acid. The compound behaves as a weak acid (pK_1 = 3.8 ± 0.2 and pK_2 = 6.8 ± 0.3 [365] and shows a wide band between 300 and 370 nm (ε = 2,325 – 2,522).

Poddar and Poddar [366], and, Chatterjee and Syamal [161] prepared $Ag\,pipz(bigH)_2]X_3]X_3$ (X = $0.5SO_4^{2-}$, OH or NO_3). Contrary to a previous report [161], the crystals exhibit no photolytic decomposition when coated with mineral oil [23].

Silver(III) compounds of bigH and substituted biguanides, 2-guanidinobenzimidazole, and biuret are also known [367 – 373]. In alkaline, neutral, or weakly acidic media (*p*H > 5), the bigH molecule possesses filled π-type delocalized orbitals. In order to form silver(III) complexes, some to the *d*-electrons of silver(I) must be forced to occupy higher anti-bonding orbitals from which these electrons may be removed by peroxydisulphate oxidation.

In acidic media, $[Ag\,en(bigH)_2]^{3+}$ undergoes a base-catalyzed intramolecular redox decomposition, leading ultimately to Ag^+ ion and the oxidation products of the ligand [33]. Contrary to an earlier report [374], $[Ag\,en(bigH)_2]^{3+}$ suffers no acid hydrolysis and is extremely stable in acidic media. This behaviour is attributed to the exceptional thermodynamic stability (log K ≈ 5.2) [375] of the complex and also to the kinetic inertness of a $4d^8$ metal ion (in a strong square field of a quadridentate ligand) towards a ligand substitution process. The $[Ag(bigH)_2]^{3+}$ ion is very similar to $[Ag\,en(bigH)_2]^{3+}$ except that the former contains one less five-membered chelate ring and possess a more open ring structure. The former compound is less stable than the latter. The *p*Ka values of $[Ag(bigH)_2]^{3+}$ and $[Ag\,en(bigH)_2]^{3+}$ are 5.9 and 3.8, respectively, which indicates that the Ag^{III}–N bonding in $[Ag(bigH)_2]^{3+}$ is comparatively weaker than that in $[Ag\,en(bigH)_2]^{3+}$. In perchloric acid media, $[Ag(bigH)_2]^{3+}$ undergoes intramolecular redox decomposition, ultimately forming Ag^+ ion and oxidation products of the ligand (Fig. 98):

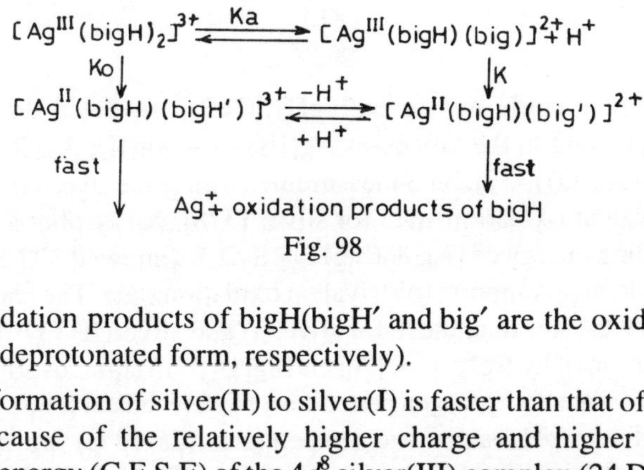

Fig. 98

Ag^+ + oxidation products of bigH(bigH′ and big′ are the oxidized form of bigH and its deprotonated form, respectively).

The transformation of silver(II) to silver(I) is faster than that of silver(III) to silver(II) because of the relatively higher charge and higher crystal field stabilisation energy (C.F.S.E) of the $4d^8$ silver(III) complex (24 Dq) compared to the corresponding silver(II) complex (12.3 Dq). Because of the high C.F.S.E., these silver(III) species are inert to ligand substitution processes. Therefore, electron transfer occurs intramolecularly without any prior ligand dissociation, as shown by the decomposition of the ligand. The $[Pd\,en(bigH)_2]^{2+}$ and $[Pd(bigH)_2]^{2+}$ ions ($4d^8$ system) aquate more slowly in an acidic medium than the corresponding nickel(II) ($3d^8$ system) [34, 249, 261]. The higher charge and smaller size of silver(III) make its complexes more inert towards ligand substitution processes than those of palladium(II).

The Ag(I) catalyzed oxidation of hypophosphorus acid by $[Ag\,en(bigH)_2]^{3+}$ quantitatively produces phosphite, Ag^+, and free ligand. Two electrons are transferred consecutively from the phosphorus(I) of the hypophosphorus acid to Ag(III) via Ag(I), instead of directly, because the complex $[Ag\,en(bigH)_2]^{3+}$ is a weaker oxidizing agent than Ag^+ [376] (Fig. 99). (XL)

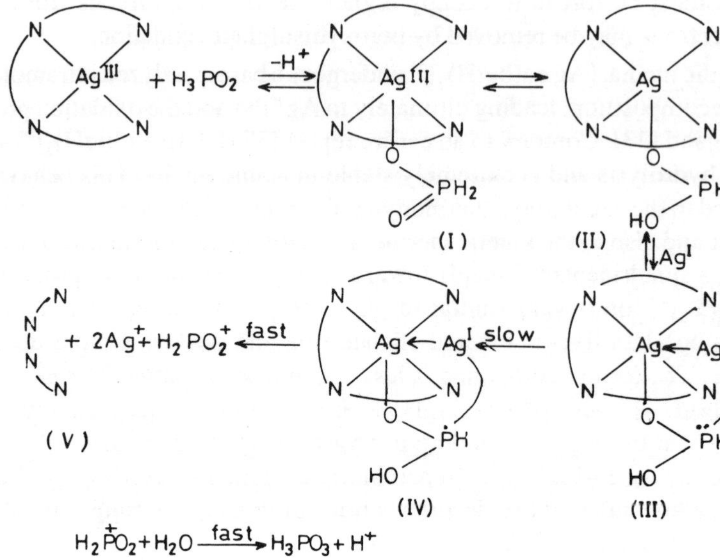

(XL)
Fig. 99

A polarographic study of $[Ag\ en(bigH)_2](NO_3)_3$ revealed three reduction waves corresponding to the processes Ag(III) ⟶ Ag(II); Ag(II) ⟶ Ag(I); and Ag(I) ⟶ Ag(O) [377]. XPS measurements on some silver(III) complexes revealed a trivalent oxidation state for silver [378]. X-ray photoelectron data on the 3*d*-binding energy of $[Ag\ en(bigH)_2](S_2O_8)_{1.5}$ showed 377.3 ev for $3d_{3/2}$ and 371.4 ev for $3d_{5/2}$, supporting a trivalent oxidation state. The chemical shifts are some 3.0 ev greater than those for silver(I) and silver(II) [379].

Reinvestigation of hydrolysis of $[Ag\ en(bigH)_2]^{3+}$ in aqueous solution shows the oxidation of water to hydrogen peroxide or the OH radical [380]. The solution kinetics at different *p*H values and temperatures (25-40°C) are first order in complex concentration and show an inverse H^+ effect and Ag^+ catalysis. At constant *p*H, the observed rate law is $\frac{-d\,(\text{complex})}{dt} = (K_A + K_B\,[Ag^+])$ [Complex]. E_a values of 3.7 and 15.2 kcal mol^{-1} have been reported for the hydrolysis of $[Ag\ en(bigH)_2](NO_3)_3$ and $[Ag\ en(bigH)_2](HSO_4)(NO_3)_2$, respectively.

The kinetics of the oxidation of $[Ag\ en(bigH)_2]^+$ to the silver(III) complex by $S_2O_8^{2-}$ consists of two consecutive pseudo first-order reactions in $[Ag\ en(bigH)_2]^+$ at excess $[S_2O_8^{2-}]$ [381]. The orders of the reaction in $[S_2O_8^{2-}]$ for the two reactants are both 1.0. For the two consecutive reactions (A) and (B) (Fig. 99) the activation energies are 13.1 ± 0.7 and 14.3 ± 3.8 kcal mol^{-1} and the values for $\log_{10}^{A}$ are 7.7 and 9.6, respectively. The mechanism which conforms to the rate is

$$[\text{Ag en(bigH)}_2]^+ \xrightarrow{S_2O_8^{2-}} [\text{Ag en(bigH)}_2]^{2+} + X \quad (A)$$

$$[\text{Ag en(bigH)}_2]^{2+} \xrightarrow{S_2O_8^{2-}} [\text{Ag en(bigH)}_2]^{3+} + X \quad (B)$$

(X may be $S_2O_8^{2-}$ or both HSO_4^- and SO_4^{2-})

Fig. 100

X does not react with $[\text{Ag en(bigH)}_2]^+$, but could probably oxidise water.

Banerjee and Banerjea [33] criticized the work of Sen *et al* [375]. The first order plots described by them are highly curved during the initial stages of the reaction and suggest a multi-step reaction. In spectral studies, the authors investigated the decomposition of $[\text{Ag en(bigH)}_2](NO_3)_3$ in perchloric acid again, where the reaction occurs in two steps—each presumably involving an intramolecular one-electron transfer process. For both steps, the rate shows an inverse dependence on acid concentration because the conjugate base of the complex is the reactive species. Despite its thermodynamic stability and acid-hydrolysis-resistant behaviour, $[\text{Ag en(bigH)}_2]^{3+}$ in the presence of silver(I) and formic acid is quantitatively reduced to Ag^+ and the free ligand [364] (Fig. 101).

$$[\text{Ag en(bigH)}_2]^{3+} + \text{HCOOH} \longrightarrow Ag^+ + CO_2\uparrow + \text{en(bigH)}_2 + 2H^+$$

Fig. 101

The high positive charge of the central metal causes an easy loss of a proton, probably from the terminal amino groups. This decomposition causes a marked change in the electronic spectrum, which is evidenced by a colour change from yellow to red when the medium is made basic. This suggests that above 0.05 *M* [H^+], the complex exists almost exclusively (>99%) in the undissociated form, confirmed by the electronic spectrum (λ_{max} = 380 nm, ε = 1,175). In solution, $[\text{Ag en(bigH)}_2]^{3+}$ is a weak oxidizing agent ($E_{1/2}$ of the Ag^{III}/Ag^{II} couple = 0.05 v); it slowly oxidizes formic acid, hydrogen peroxide and alcohols (ethyl alcohol, isopropanol, butanol) to carbon dioxide, oxygen, and carbonyl compounds, respectively, while the complex cation is reduced to Ag^+ and the free ligand. The authors [364, 382, 383] proposed the following empirical rate law (Fig. 102):

$$\frac{-d[\text{Ag en(bigH)}_2]^{3+}}{dt} = \frac{k_1\,[\text{HCOOH}]\,[Ag^+]^2}{[H^+]} \text{ (for HCOOH)}$$

$$= \frac{k_2 + k_3\,[Ag^+]\,[H_2O_2]\,[\text{complex}]}{[H^+]} \text{ (for } H_2O_2)$$

$$= \left(k_4 + \frac{k_5}{[H^+]}\right) [\text{ROH}]\,[\text{complex}] \text{ (for alcohols)}$$

Fig. 102

Recently Banerjee [384] studied the kinetics of oxidation of eleven organic and inorganic substrates by the $[Ag\ en(bigH)_2]^{3+}$ ion in aqueous perchlorate media. Widely different rates and mechanistic versatility were noticed, depending upon the nature of the reductant, inner sphere, outer sphere, etc. The Ag^+-catalysis may involve a Ag^+–Ag° or Ag^I/Ag^{II} catalytic cycle. The silver(III) complex is kinetically more active than its conjugate bases towards a neutral or anionic reductant. Towards a cationic reductant, however, the conjugate bases are more active. This complex exhibits some kinetic selectivity in its reactions and could therefore be utilised in kinetic methods for the estimation of benzyl alcohol and hypophosphite.

Oxidation of silver(I) salts by peroxydisulphate in the presence of *p*-phenetyl biguanide sulphate ($LH.H_2SO_4$) at *p*H ~ 5-6.5, yields two varieties of silver(II) complexes:

α-$[Ag^{II}(LH)_2]SO_4.H_2O$ (yellow-brown) and β-$[Ag^{II}(LH)_2]SO_4.H_2O$ (black).

The corresponding nitrate formed only one type (black). Their acid solutions liberated one electron equivalent of iodine from potassium iodide solution. These complexes show an absorption band at ~25,000 cm^{-1} (reflectance spectra), corresponding to a *d-d* transition. The effective magnetic moments are 1.90 B.M. (yellow) and 1.85 B.M. (brown) [370].

When silver nitrate in MeOH is added to an EtOH solution of $[Ni(dichloro\text{-}bigH)_2]Cl_2$ or $[Ni(dichloro\text{-}R\text{-}bigH)_2]Cl_2.nH_2O$, white crystals of $[Ag(H_2O)_2(dichloro\text{-}bigH)]Cl.H_2O$ separate. The complexes are tetrahedral [37], and some are dimeric [368]. From conductometric titration data, Gupta and Kaushal [275] showed that paludH combines with $AgNO_3$ to yield a 1 : 2 complex. A comparison of the IR spectra of silver(I) and silver(III) biguanides indicates the presence of stronger metal-ligand bonding in the silver(III) complexes.

(xiii) Molybdenum(VI), (V), and (III)

By the reaction of bigH and substituted bigH (Et-bigH or Ph-bigH) with 12-molybdophosphoric acid, several salts of 12-molybdophosphoric acid have been synthesized. They are soluble in water, and the dry crystalline compounds contain large amounts of water of crystallization. Complete loss of this water takes place at ~ 250°C; when the salts are heated above 450°C, they begin to decompose [385].

Preparation of some molybdenum(V) oxo species with substituted biguanides (Ph-bigH, *p*-Cl-bigH, Ph-bigH, N^1-*p*-Cl-Ph-N^5-*iso*propyl-bigH, Me_2-bigH, morph-bigH, or paludH) of the following types has been reported [386-391]: $(R\text{-}bigH_2)_2[Mo_2O_4Cl_4(H_2O)_2]$, $[Mo_2O_3(R\text{-}big)_4]$, $(R\text{-}bigH_2)_2[Mo_2O_4(C_2O_4)_2(H_2O)_2]$, $[Mo_2O_4(H_2O)_2(R\text{-}big)_2]$ and $(R\text{-}bigH_2)$-$[MoOCl_4(H_2O)]$. The complexes exhibit two absorption bands at 22,600 cm^{-1} and 14,000 cm^{-1}, which are assigned to $b_2 \longrightarrow b_1$ and $b_2 \longrightarrow e$ transitions, respectively. Their bonding parameters were evaluated by EPR studies. The existence of a monomer-dimer equilibrium has been noted; in DMF the equilibrium is displaced towards the dimeric form (Fig. 103).

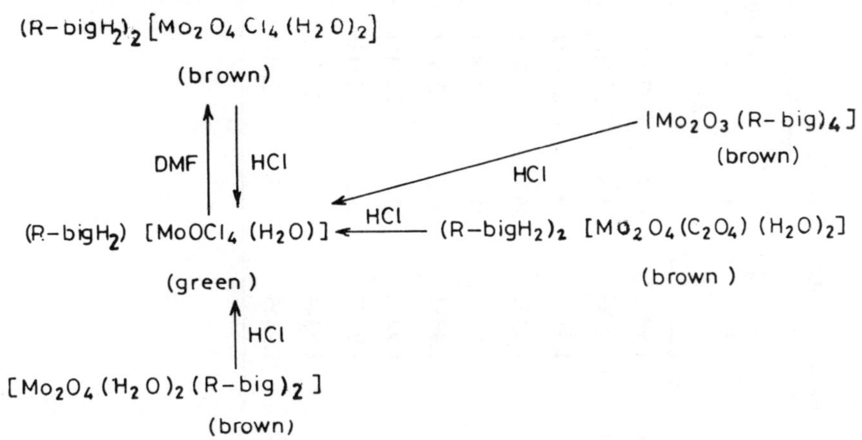

Fig. 103

The compounds exhibit intense bands at 950-975 cm^{-1}, characteristic of Mo = O bonding. The dimeric compounds exhibit vibrations at 620 cm^{-1} and 730 cm^{-1}, corresponding to a 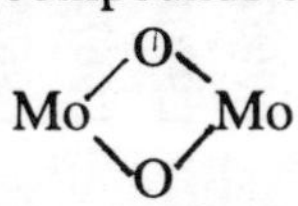 bridge. Compounds of the type $Mo_2O_3(R\text{-}big)_4$ show a band at ~860-880 cm^{-1}, which may be assigned to a simple oxygen bridge, Mo-O-Mo. The EPR parameters are shown in Table 19 [391] :

Ammonium hexathiocyanatomolybdnate(III), $(NH_4)_3[Mo(NCS)_6]$, when treated in absolute methanol with substituted biguanides (R-bigH = Me-bigH, Me_2-bigH or Ph-bigH) yields compounds of the type $[Mo(R\text{-}bigH)_3](NCS)_3$. Stable compounds, $[Mo(NCS)_2\,en(bigH)_2]NCS$ and $[Mo(NCS)_2\,pipz(bigH)_2]$ NCS were also isolated with $en(bigH)_2$ or $pipz(bigH)_2$, respectively. These compounds undergo hydrolysis in water, but the rate for $[Mo(bigH)_3](NCS)_3$ is considerably less than that of $[Mo(en)_3](NCS)_3$, which may be explained in terms of the delocalized π-electron system of metal-biguanide complexes [78].

The conductance values of $[Mo(R\text{-}bigH)_3](NCS)_3$ type complexes in methanol (245-252 ohm^{-1} cm^2 mol^{-1}) indicate tri-univalent electrolytes, while that of $[Mo(NCS)_2\,en(bigH)_2]NCS$ in methanol (75.8 ohm^{-1} cm^2 mol^{-1}) indicates a 1 : 1 electrolyte. The compounds are paramagnetic (μ_{eff} = 3.68-3.74 B.M.), supporting the presence of three unpaired electrons, and they show a *d-d* band at about 23,530-26,040 cm^{-1}. Comparison of the electronic spectra of $[Mo(en)_3](NCS)_3$ and $[Mo(bigH)_3](NCS)_3$ suggests that the ligand field strength of bigH towards molybdenum(III) is almost equal to that of en. X-ray

Table 19 g-Values and Hyperfine Structure Constants for Molybdenum(V) Complexes [391]

Complex	*Medium*	g_{iso}	α_{iso}	$g_{\parallel}$	$A_{\parallel}$	$g_{\perp}$	$A_{\perp}$	ΔE *in* cm^{-1}	
								$b_2 \rightarrow b_1$	$b_2 \rightarrow e$
$(Me_2\text{-bigH}_2)[MoOCl_4(H_2O)]$	DMF	1.9458	51.9	1.9669	82.4	1.9353	36.7	22,600	14,000
	HCl	1.9490	51.5	1.9698	82.9	1.9386	35.8	22,600	14,000
$(\text{morph-bigH}_2)[MoOCl_4(H_2O)]$	DMF	1.9453	51.7	1.9658	82.7	1.9350	36.2	22,600	14,000
	HCl	1.9464	51.7	1.9689	82.8	1.9352	36.1	22,600	14,000
$(\text{Ph-bigH}_2)[MoOCl_4(H_2O)]$	DMF	1.9451	51.9	1.9682	82.2	1.9333	36.7	22,600	14,050
	HCl	1.9488	51.7	1.9701	82.1	1.9381	36.5	22,600	14,000
$(\text{Cl-Ph-bigH}_2)[MoOCl_4\ (H_2O)]$	DMF	1.9455	52.1	1.9673	82.9	1.9346	36.7	22,600	14,050
	HCl	1.9478	51.1	1.9695	82.9	1.9370	35.2	22,600	14,000
$(\text{paludH}_2)[MoOCl_4\ (H_2O)]$	DMF	1.9462	51.6	1.9676	82.8	1.9344	36.3	21,700	13,800
	HCl	1.9474	55.4	1.9705	83.3	1.9359	41.5	22,500	14,000

powder photographs show $[Mo(bigH)_3]^{3+}$, $[Cr(bigH)_3]^{3+}$, and $[Co(bigH)_3]^{3+}$ to be isomorphous. The IR spectra of thiocyanato complexes support Mo-NCS bonding. The ν (CN) and ν (CS) stretches of $[Mo(NCS)_2\,en(bigH)_2]NCS$ were found at 2,090 cm^{-1} (s) and 785 cm^{-1} (w).

(xiv) Oxovanadium(IV)

Oxovanadium(IV) complexes of biguanides (R-bigH = alkyl biguanides, $en(bigH)_2$, $trime(bigH)_2$, $hex(bigH)_2$ or $pipz(bigH)_2$) have been prepared by the action of alkali on a mixture of vanadyl sulphate and a large excess of the ligand [1, 14]. During T.G. studies of $[VO(H_2O)(big)_2]$ Babykutty *et al* [208] observed that coordinated water molecules are lost at 160°C, forming pentacoordinated $[VO(big)_2]$, which is totally decomposed to vanadium pentoxide above 450°C. All the compounds are hydrolyzed by acids, are paramagnetic (μ_{eff} = ~ 1.7 B.M.) and resemble $[VO(AAU)_2]$ [5].

Complexes of the types $[VO(R\text{-}big)_2]$, $[VO(H_2O)\,(R\text{-}big)_2]$, $[VO(H_2O)\,en(big)_2]$, $[VO(H_2O)\,trien(big)_2]$ and $[VO(H_2O)\,pipz(big)_2]$ exhibit a ν (V = O) stretch at 940 cm^{-1} [392], a remarkably lower value than those for the other oxovanadium (IV) complexes. This is consistent with strong out-of-plane vanadium-biguanide $d_\pi \rightarrow p_\pi$ bonding, which can account for the decrease in ν (V = O) stretch. There is a shift of ν (V = O) stretching frequency from bigH to R-bigH oxovanadium(IV) complexes, which may be related to the dissociation constants of the free ligands. All the compounds show three electronic spectral bands around 16,700 cm^{-1}, 23,200 cm^{-1}, and 25,000 cm^{-1}, due to $d_{xy} \rightarrow d_{xz}, d_{yz}$; $d_{xy} \rightarrow d_{x^2-y^2}$, and $d_{xy} \rightarrow d_{z^2}$ transitions, respectively. The red shift of the bands occurs in dilute acetic acid solutions of the complexes, indicating decomposition of the complexes [1, 14, 162].

The electronic spectra and EPR parameters of some oxovanadium(IV) complexes (Table 20) show that the bonding of all these complexes is very similar, a fact confirmed by the constant dipolar term $|A_\perp - A_\parallel|$. Comparison of the average hyperfine splitting $\langle A \rangle$ for oxovanadium(IV) complexes confirms that bigH is also strong field in character [14, 224].

(xv) Manganese(IV), (III), and (II)

Addition of aqueous potassium permanganate to a strongly basic solution of $bigH.H_2SO_4$ produces red crystals of dihydroxobis(biguanide)manganese(IV) hydroxide, $[Mn(OH)_2(bigH)_2](OH)_2$. This compound was also prepared by oxidation of a manganese(II) salt with either hydrogen peroxide or sodium per oxydisulphate, followed by addition of a basic solution of $bigH.H_2SO_4$. Treatment with dilute acids, (nitric acid, acetic acid etc.) yields the corresponding salts. Reddish brown crystals of dihydroxo(ethylenedibiguanide)manganese(IV) hydroxide dihydrate, $[Mn(OH)_2\,en(bigH)_2](OH)_2.2H_2O$, have also been reported. Addition of aqueous potassium permanganate to a strongly basic solution of hex(bigH)$_2$.2H_2SO_4 yields [Mn(OH)$_2$ hex(bigH)$_2$](OH)$_2$.

The magnetic moment of dihydroxomanganese(IV) biguanide complexes (2.0-2.83 B.M. at 30°C) is much lower than the value of 3.87 B.M. for manganese(IV) [1, 14]. The ligand field spectra of [Mn(OH)$_2$(bigH)$_2$](OH)$_2$ exhibit three absorption bands at 19,230 cm^{-1} ($^4A_{2g} \rightarrow {}^4T_{1g}$), 17,856 cm^{-1} ($^4A_{2g} \rightarrow {}^2T_{1g}$) and 15,923 cm^{-1} ($^4A_{2g} \rightarrow {}^4T_{2g}$), with the middle band being much less intense [167]. The authors also calculated Dq = 1,592 cm^{-1}, and the usual ligand field parameters F_2 = 1,007 cm^{-1} and F_4 = 146 cm^{-1} [167].

Cryomagnetic studies of some manganese(IV) compounds, viz., [MnO(big)$_2$]$_2$.6H_2O, [MnO(bigH)$_2$]$_2$(NO$_3$)$_4$.2H_2O and [MnO(bigH)$_2$]$_2$-(SO$_4$)$_2$.11H_2O, show a gradual decrease of magnetic moment with decreasing temperature for all the compounds, indicating the presence of antiferromagnetic interaction [168]. The maxima at 290 K and 250 K in the atomic susceptibility vs. temperature curves for [MnO(big)$_2$]$_2$.6H_2O and [MnO(bigH)$_2$]$_2$(NO$_3$)$_4$.2H_2O may be due to the Néel temperature. The higher calculated values of atomic susceptibilities compared with the experimental values may be attributed to the variation of exchange energy (– 2J) with temperature. The high exchange integral values (62.6 cm^{-1}, 59.1 cm^{-1}, and 66.0 cm^{-1} for the hydrate, nitrate, and sulphate salts) suggest strong super-exchange interaction and metal-metal direct interactions (XLI) (Fig. 104).

The electron density initially on the t_{2g} orbitals of one metal ion is transferred to the other metal ion through the 2*p* orbitals of the bridging oxygen atoms, causing a strong exchange interaction. The direct metal-metal interaction occurs through the normalized overlap of the *d* orbitals of the metal ion.

A polarographic study of [Mn(OH)$_2$(bigH)$_2$](OH)$_2$ shows three polarographic waves—the first step for the reduction of manganese(IV) is a two-electron step giving manganese(II) with a half-wave potential almost equal to that for the reduction of manganese(III) to manganese(II) [393].

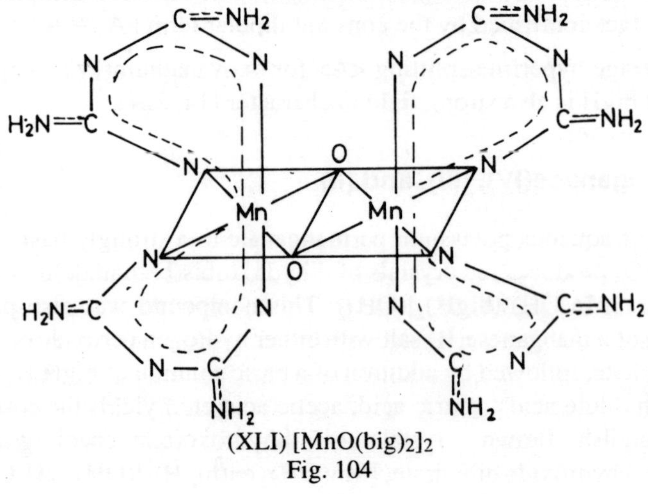

(XLI) [MnO(big)$_2$]$_2$

Fig. 104

Table 20 EPR Data for Oxovanadium(IV) Complexes [162]

Complex	*Medium*	g_{av}^{*}	$g_{\perp}$	$g_{\parallel}$	$\langle A\rangle$**	$A_{\perp}$	$A_{\parallel}$
					(10^{-4} cm^{-1})		
$[VO(H_2O)(big)_2]$	Ni(II) complex	1.97	1.99	1.94	87	54	154
	25%(v/v) CH_3COOH–H_2O	1.97	–	–	–	–	–
$[VO(H_2O)$ en$(big)_2]$	Ni(II) complex	1.97	1.99	1.94	88	55	153
$[VO(H_2O)$ trime$(big)_2]$	Ni(II) complex	1.97	1.99	1.94	88	56	152
$[VO(H_2O)$ tetme$(big)_2]$	Ni(II) complex	1.97	1.99	1.94	87	53	155
$[VO(H_2O)$ hex$(big)_2]$	Ni(II) complex	1.97	1.99	1.94	87	53	155
$[VO(H_2O)$ pipz$(big)_2]$	Ni(II) complex	1.97	1.99	1.94	87	50	154
$[VO(H_2O)(AMU)_2]$	Ni(II) complex 25% (v/v) CH_3COOH–H_2O	1.97	–	–	101	–	–

*$g_{av} = 1/3(g_{\parallel} + 2g_{\perp})$ except for $[VO(H_2O)(L)_2]$ (L = big or AMU) in dilute acetic acid

**$\langle A\rangle = 1/3\ (A_{\parallel} + 2A_{\perp})$ except for spectra in dilute acetic acid

Dark chocolate-coloured glistening crystals of tetrakis(biguanide)-di-μ-hydroxodimanganese(III), $[(big)_2Mn(OH)_2Mn(big)_2].H_2O$, were prepared either (1) by air-oxidation of manganous sulphate and a strongly basic solution of $bigH.H_2SO_4$ or (2) by adding bigH to freshly prepared manganese(III) acetate in basic medium. With $en(bigH)_2$, $[Mn(H_2O)(OH)\ en(big)_2].0.5H_2O$ was deposited. Addition of acetic acid to $[Mn(H_2O)(OH)\ en(big)_2].0.5H_2O$ yields a deposit of $[Mn(H_2O)(OH)\ en(big)_2](CH_3COO)_3$. With the exception of $[Mn(H_2O)(OH)\ en(big)_2].0.5H_2O$, the compounds show normal magnetic moment values ($\mu_{eff} \sim 4.9$ B.M.) [394]. The lower value of 3.22 B.M. for $[Mn(H_2O)(OH)(big)_2].0.5H_2O$ suggests that the compound must be regarded as an inner-level complex with d^2sp^3 octahedral bonds but with an incomplete quenching of its orbital moment.

Poddar and co-workers [395] refluxed $[Mn(acac)_3]$ and the free base in dry EtOH and obtained dark chocolate-coloured $[(big)Mn(OH)_2Mn(big)].H_2O$. The formation of $[Mn(big)_3]$ was prevented by the presence of water. Refluxing $[Mn(acac)_3]$ and the free big (1 : 2) in EtOH yields red crystals of $[Mn(OH)(H_2O)(acac)(big)]$. The conductance values (38.7 $ohm^{-1}\ cm^2\ mole^{-1}$ in MeOH at 32°C) and weak alkalinity of the MeOH solution indicate that the heterochelate is protonated by water in the solvent (Fig. 105):

$$[Mn(OH)(H_2O)(acac)(big)] + H_2O \longrightarrow [Mn(OH)(H_2O)(acac)(bigH)]OH$$

Fig. 105

The interaction of tris(benzoylacetone)manganese(III) with big in a 1 : 2 ratio in EtOH produces the heterochelate (benzoylacetone)bis(big)manganese(III) [396].

Dey *et al* [397] reported the preparation of hydroxoaquapiperazinedi-biguanidemanganese(III), $[Mn(OH)(H_2O)\ pipz(big)_2]$, whose magnetic moment value ($\mu_{eff} = 2.88$ B.M.) suggests that it is a low-spin manganese(III) complex. Poddar [245] prepared a manganese(II) complex of glycosyl-biguanide, $[Mn(H_2O)(glycosylbig)]$. Other biguanides yield similar types of compounds.

(xvi) Rhenium(V)

Rose-violet crystals of oxohydroxorhenium(V)bis(biguanide)hydroxide, $[ReO(OH)(bigH)_2](OH)_2$, were prepared by Ray *et al* [398] by treating strongly alkaline $bigH.H_2SO_4$ solution with potassium hexachlororhenate(IV) in the presence of air. On treatment ammonium chloride $[ReO(OH)(big)_2]$ yielded a yellow precipitate of $[ReO(Cl)(big)_2]$. Addition of dilute acid to the complex base yields the salts $[ReO_2(bigH)_2]X$ (X = Cl, Br, I, SCN, NO_3, CH_3COO, 0.5 C_2O_4, $0.5CO_3$, $0.5SO_4$, or $0.5HPO_4$). Such salt formation indicates that the base undergoes a change owing to the transfer of a proton from the complex to the outer zone, yielding $[ReO_2(big)_2]OH.H_2O$. Both the salt and base compounds are diamagnetic, suggesting Re-Re bond formation in the complexes [1].

Rhenium(V) in combination with bigH has a pronounced tendency to form multiple bonds with oxygen. Ray [399] prepared a series of complexes containing such bonds of the type [ReO(OH)(bigH)](OH)X (X = OH, CH_3COO, Cl, 0.5 SO_4, or bigH in the zwitterion form).

Ray also discussed their configuration, magnetic properties, and the nature of the linkages between the metal and coordination sites of the ligand via IR spectra [399]. Recently Roychowdhury *et al* [400] prepared and characterized $[ReO(OH)(big)_2]$, $[ReOCl(big)_2]$. $[ReOCl(bigH)_2]Cl_2$, $[ReCl_2(bigH)_2]Cl_3$, $[Re(OH)_2(bigH)_2]Cl_3$, $[ReO(OH)(Me\text{-}big)_2]$, $[ReOCl(Me\text{-}big)_2]$, $[ReOCl(Me\text{-}bigH)_2]Cl_2$, $[ReCl_2(Me\text{-}bigH)_2]Cl_3$, and $[Re(bigH)_3](SO_4)_{1.5}.7H_2O$.

Treatment of $[ReO(OH)(big)_2]$ with ammonium chloride yields $[ReOCl(big)_2]$, but reaction with dilute hydrochloric acid produces $[ReOCl(bigH)_2]Cl_2$, which contains protonated biguanide. Treating the same compound with concentrated hydrochloric acid in acetone produces $[ReCl_2(bigH)_2]Cl_3$. The neutral compounds are insoluble in water, whereas the ionic compounds are soluble in water, methanol, and ethanol. All the compounds are diamagnetic except for $[Re(bigH)_3](SO_4)_{1.5}.7H_2O$ (μ_{eff} = 0.74 B.M.) whose paramagnetism may be due to its dinuclear nature involving SO_4^{2-} bridging and Re-Re interaction. Diamagnetism caused by major complexes is due to tetragonal distortion (caused by non-equivalent ligands) of the octahedral field and a small distortion due to nonuniform occupancy of the t_{2g} level. Moreover, d_π–p_π bonding in metal-oxygen is stronger than in metal-biguanide, further destabilizing the t_{2g} level, making the d_{xy} orbital the lowest energy orbital and accommodating the two electrons with their spins paired.

The ν (Re = O) bands for oxo complexes $[ReOCl(L)_2]$ and $[ReOCl(LH)_2]Cl_2$ appear at 990-980 cm^{-1}. The medium intensity ν (Re-Cl) bands appear at 935-920 cm^{-1} in the chloro and oxochloro complexes. The band at 920 cm^{-1} in the hydroxo complexes is attributed to ν (Re-OH), and the bands at 750 cm^{-1} arise from Re-OH deformation modes. The strong band at 490 cm^{-1} is assigned to ν (Re-N).

(xvii) Osmium(VIII), (VI), (IV), (III), and (II)

Yellow, crystalline osmium (VIII) complexes of bigH and AUH of the types $[(O)OsN(bigH)_2]X_3$ and $[(O)OsN(AU)_2]X$ (X = OH or 0.5SO_4) were prepared by adding the ligand normal salt to an aqueous solution of potassium osmiamate at a *p*H > 10.

The TGA curves for the complexes indicate that the lattice water is lost at 80-100°C, and the anhydrous compound thus formed begins to decompose at ~ 350°C. The weight of the residue left at ~ 400°C is less than that required for osmium(IV) oxide, indicating some loss of osmium as osmium tetroxide. The Os ≡ N and Os = O peaks occur at ~ 1,025 cm^{-1} and 800-900 cm^{-1}, respectively (Table 21) [258, 401, 402].

Addition of a strongly alkaline solution of bigH.H_2SO_4 to an acidic solution of OsO_4, followed by digestion of the mixture yields yellow, crystalline osmylbis(biguanide), [OsO_2(big)$_2$]. This compound undergoes no change in weight even on heating to 130°C for several hours. On dissolving in concentrated hydrochloric acid, it yields an olive-green solution, from which addition of bigH.H_2SO_4 deposits olive-green crystals of osmylbis(biguanide)sulphate, [OsO_2(bigH)$_2$]SO_4.H_2O [403]. Compounds of the types

[OsO_2(bigH)$_2$]Cl_2,[OsO_2(big)(γ-pic)$_2$] and [OsO_2(big)(dipy)]Cl

were also prepared and characterized. The compounds are diamagnetic and liberate two equivalents of iodine per osmium atom, indicating osmium(VI) → osmium(IV). IR spectra of the sulphate salts indicate that the two oxo groups are *trans* to each other [403].

With substituted biguanides (LH = Me_2–bigH, morph–bigH or paludH) osmium(IV) compounds of the type [LH_2]$_2$[$OsCl_6$] were reported [404]. The [LH_2]$_2$[$OsBr_6$] compounds are very stable, being unaffected by cold and hot hydrohalogen acids. Heating (Me_2–bigH$_2$)$_2$[$OsCl_6$] with metallic osmium yields violet [Os(Me_2-bigH)$_3$]$^{2+}$. Heating (LH_2)$_2$[OsX_6] (X = Cl or Br) to above 120°C causes elimination of the hydrohalogen acids and yields a mixture of [Os(LH)$_2X_2$] and [Os(L)$_2X_2$]. The authors also discussed the ultraviolet and visible spectra of the complexes [163, 404].

By reduction of a mixture of $(NH_4)_2$[$OsBr_6$] and R-bigH.HCl with metallic osmium in aqueous solution [Os(R-bigH)$_3$]Br_2 compounds are formed. The deep violet aqueous solutions, on treatment with various oxidants (Ce^{4+}, $S_2O_8^{2-}$, H_2O_2, or ClO_2^-), become green and then light yellow (Fig. 106):

$$\underset{\text{violet}}{\overset{\text{II}}{[\text{Os(R-bigH)}_3]^{2+}}} \longrightarrow \underset{\text{green}}{\overset{\text{III}}{[\text{Os(R-bigH)}_3]^{3+}}} \longrightarrow \underset{\text{yellow}}{\text{Os(IV)}}$$

Fig. 106

Synthesis of other osmium(II) and osmium(III) complexes of the types [Os^{II}(L)$_2$(R-bigH)$_2$], [Os^{II}(R-bigH)$_3$]X_2, [Os^{III}(R-bigH)$_3$]X_3, [Os^{III}(L)$_2$(R-bigH)$_2$]X, [Os^{III}(py)$_2$(paludH)$_2$]X_3 and [Os^{III}(en)(paludH)$_2$]X_3 (R-bigH = Me_2-bigH, morph-bigH, *p*-Cl-Ph-bigH, Pr^i-bigH or paludH (L = Cl or Br; X = NCO, I, SCN, N_3, BPh_4, H_2PO_2, 0.5S_2O_3, 0.5SO_4 or 0.5 [Fe(CN)$_5$(NO)]) are also known. [Os^{II}(paludH)$_3$]($SbOC_4H_4O_6$)$_2$ has been resolved into optical isomers (*l*-[Os^{II}(paludH)$_3$($SbOC_4H_4O_6$)$_2$, $[\alpha]_D$ = (–)180°, $[M)_D$ = (–)2169°; *d*-[Os^{II}(paludH)$_3$]($SbOC_4H_4O_6$)$_2$, $[\alpha]_D$ = (+)190°, $[M]_D$ = (+)2289°) [163, 403-405].

Osmium(II) complexes are diamagnetic, whereas osmium(III) complexes have magnetic moments of 1.60-1.68 B.M. [163, 404, 405]. The absorption spectra of tris(biguanide) osmium(II)/osmium(III) (Table 21) indicate the transfer of intramolecular energy from the π-electron system of the ligand to the 5*d* electron system of osmium. Tris(ligand)osmium(III) compounds exhibit similar absorption spectra, consisting of a strong ligand band $\pi \rightarrow \pi^*$ between 50,000-30,000 cm^{-1} and two marked charge transfer bands—one at 25,000 cm^{-1}, log ε = ~ 4 and the other at 18,800 cm^{-1}, log ε = ~ 3.8 [163].

(xviii) Iron(III) and (II)

Light greenish-brown iron(III) complexes of the types $[FeCl(big)_2]Cl$, $[FeCl(Ph\text{-}big)_2]Cl$, $[FeCl(p\text{-}tolyl\text{-}big)_2]Cl$, and $[FeCl(o\text{-}Cl\text{-}Ph\text{-}big)_2]Cl$ were prepared by adding anhydrous $FeCl_3$ to a clear solution of bigH/subst-bigH base in dry acetone [406].

Temperature variation in magnetic susceptibilities between 300 K and 77 K are in accord with the spin quartet (S = 3/2) ground state of iron(III). Isomer shifts with respect to a Cu/ ^{57}Co source and quadrupole splitting in the Mössbauer spectrum at room temperature are 0.205 and 0.75 mm/s, respectively [406].

Iron(II) reacts with bigH/Ph-bigH and $en(bigH)_2$ in the presence of CN^- ion to form a red, soluble compound. Ph-bigH also reacts with iron(III) and also yields the same colour, presumably because iron(III) is reduced to iron(II) by the reagent. However, no red colour appears in the presence of oxidizing agents [54, 407, 408]. The colour is due to charge transfer ($t_{2g} \rightarrow \pi^*$) from iron to Ph-bigH, based on high molar absorptivity [54].

With glucosylbiguanide and glucosylalkylbiguanides, Poddar prepared octahedral $[Fe(ligand)_2]$, with magnetic moments of 6.0-6.4 B.M. [245]. The salts react with biguanidebenzohydroxamic acid at *p*H 2-5 and give a clear violet solution (ν_{max} = 22,730-23,810 cm^{-1}). Unfortunately, the solid compound could not be isolated [203].

Treatment of free bigH base with tris(acetylacetone)iron(III) in an alcoholic medium yields a heterochelate $[Fe(acac)(bigH)_2](OH)_2$, which is paramagnetic (μ_{eff} = 5.9 B.M.) at room temperature [409]. The reaction of ferrous sulphate with paludH in methanol yields $[Fe(H_2O)_2(paludH)_2]SO_4$. The authors proposed a probable structure for this complex [275]. The characteristic IR bands for $C_6H_4\text{-}\overset{\oplus}{N}H$ and $-NH_2$ groups have been identified at 1,290 cm^{-1} and 1,735 ± 5 cm^{-1}. Bands at 815, 1,650, and 3,400 cm^{-1} support the coordination of aqua molecules in the metal chelates.

(xix) Boron(III)

Biguanides, guanylurea, and l-amindino-*O*-alkylureas react with trimethoxyborate and boric acid in methanol and water, respectively, yielding crystalline compounds of the types $LB(OCH_3)_2$ and $LB(OH)_2$ (L = the chelating ligand) (Fig. 107) :

These compounds are insoluble in common organic solvents and thermally stable up to 200°C. Hydrolysis reactions occur both in neutral and acidic media,

Table 21 IR Spectral Data (cm^{-1}) for bigH.H_2SO_4, AUH.HCl and Osmium-biguanide Complexes [401]

bigH.H_2SO_4	*AUH.HCl*	*[(O)OsN (bigH)$_2$](OH) SO_4*	*[(O)OsN (Me-bigH)$_2$]-(OH)SO_4.H_2O*	*[(O)OsN (Me$_2$-bigH)$_2$]-(OH)SO_4.H_2O*	*[(O)OsN (Et$_2$-bigH)$_2$]-(OH)SO_4.H_2O*	*[(O)OsN (AU)$_2$]-(OH).2H_2O*	*Assignment*
3,450 vs	3,350	3,420 s	3,410 sh	3,410-3,350 bb	3,420 sh	3,320 w	ν (NH) in \\C-NH_2 //
3,348 vs	3,320	3,350 s	3,355 w	–	–	3,300 s	—
3,190 sh	3,180	3,145 vs	3,150 s	3,150 vs	3,125 vs	3,145	ν (NH) in > C = NH
3,030 sh	–	–	3,090 s	3,100 s	–	–	—
–	1,740 vs	–	–	–	–	–	ν (C = O)
1,680 vs	1,680 vs	1,650 vs	1,645 vs	1,640 vs	1,640 vs	1,645 s	ν (N–C–N)
	1,625 w						
1,600-1,550 bb	1,580 vs	1,550 s	1,540 w	1,550 vs	1,555 w	1,530 s	δ (NH_2)
	1,530 vs						
1,470 vs	1,465 vs	1,450 bb	1,460 s	–	1,460 s	1,455 s	
–	–	1,255 vs	1,250 s	1,260 vs	1,260 s	1,250 s	ν (N–C–N) ring vibration
1,115 s	–	1,095 s	1,120 w	1,110 s	1,110 w	–	—
1,100 s	–	1,120 w	–	1,130 s	1,130 s	–	ν (SO_4)
1,047 s	–	–	1,055 s	1,055 s	1,070 w	–	—
–	–	1,030 vs	1,020 vs	1,020 vs	1,020 s	1,025 vs	ν (Os ≡ N)
–	–	{820 s 870 vs	{825 vs 870 vs	{820 vs 860 vs	{820 vs 865 vs	{825 s 880 vs	ν (Os = O)
725 s	720 w	725 w	–	720 s	725 w	720 s	NH_2-wagging

AUH = N^1-amidinoisoureas;

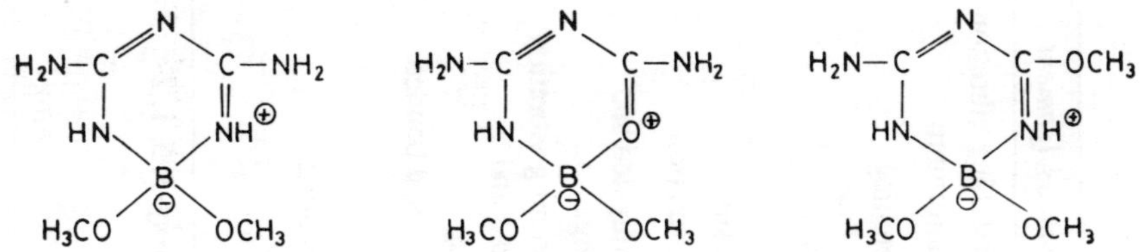

(XLII) (XLIII) (XLIV)

Fig. 107

which can be explained by considering the protonation of the N atom between the C atoms of the chelating ring distributing the π-electron delocalization and consequent aromaticity. The π-bond between the boron and nitrogen atoms thus becomes susceptible to hydrolytic attack and eventual rupture (Fig. 108).

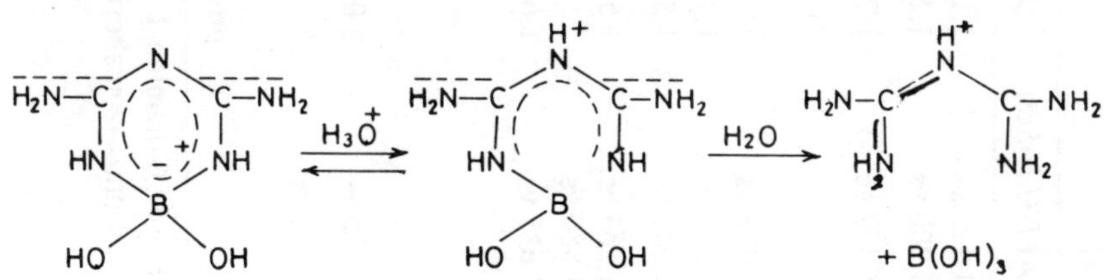

(XLV) (XLVI) (XLVII)

Fig. 108

The hydrolysis and absorption data show that the compounds possess aromaticity. The identical λ_{max} values (ultraviolet), for the free ligands as well as their boron complexes suggest little perturbation of the π-electron system of the ligands by boron. The IR spectral bands of the boron heterocyclics and their probable assignments are shown in Table 22.

Thermogravimetric analyses of these compounds indicate that they are stable up to 200°C and rapidly decompose between 250°C and 300°C. Thermal data suggest the elimination of a molecule of methanol as a first step in the decomposition process [165, 410] (Fig. 109) (XLVIII):

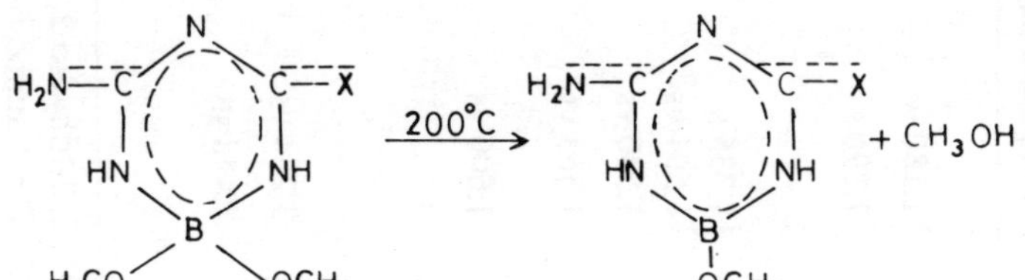

(X= NH_2, OCH_3 or OH)

(XLVIII)

Fig. 109

Table 22 IR Spectral Bands (cm^{-1}) of Boron Heterocyclics and Their Probable Assignments [165]

bigH	*Compound (A)*	*AUH.HCl*	*Compound (B)*	*AEUH.H_2SO_4*	*Compound (C)*	*Probable assignment*
				1,125 s	1,136 s	C-OR stretching vibration
1,201	1,180 w	1,200 s	1,201 s	1,200 vs	1,205 vs	NH_2-guanidinium
1,260 w	1,290 s	–	1,304 vs	1,307 sh	1,285 sh	N–H stretching vibration free and bonded
1,316 w	1,335 s	–	–	1,333 vs	–	–
	1,504 vs	–	–	–	1,492	B-N bonded
1,540 vs	1,540 vs	1,530 vs	1,540 vs	1,540 vs	1,540 vs	N-H deformation
1,575 vs	1,563 vs	1,580 s	1,564 vs 1,580 vs	1,563 br 1,580 vs	1,563 s	C–N conjugated and in the ring
1,628 vs	1,600 vs	1,625 w	1,638 vs	1,635 vs	1,600 vs	C=N stretching vibration conjugated and in the ring
–	–	1,740 vs	1,704 vs	–	–	C = O free and bonded
3,125 s	3,284 vs	3,180 s	3,179-3284	3,125 vs	3,028	Amide-I
3,284 s	3,448 sh	3,320 s	(broad)	3,284 vs	3,232	N–H stretching vibration
		3,350 s			broad	

A = 4 : 6-diamino-1, 2, 3-trihydro-2, 2-dimethoxy-3, 5, 1, 2-diazaazoniaborinide; B = 4 : 6-diamino-1, 2, 3-trihydro-2 : 2-dimethoxy-3, 5, 1, 2-diazaoxoniaborinide; C = 4-amino-6-ethoxy-1, 2, 3-trihydro-2 : 2 dimethoxy-3, 5, 1, 2-diazaazoniaborinide
vs = very strong; s = strong; w = weak; sh = shoulder; b = broad

Cationic boron(III) chelates with such ligands having the composition $B(L)_2^+ X^-$ have been synthesized by the transamination reaction between tris-(diethylamino)borane and bigH. The compounds have very stable B-N bonds and can be recrystallized from aqueous solution.

(xx) Beryllium(II)

Condensation of free base bigH with beryllium(II) acetylacetonate in dry methanol yields the Schiff base complex Be(big_2ac) (m.p. 214-216°C) (Fig. 110) (XLIX):

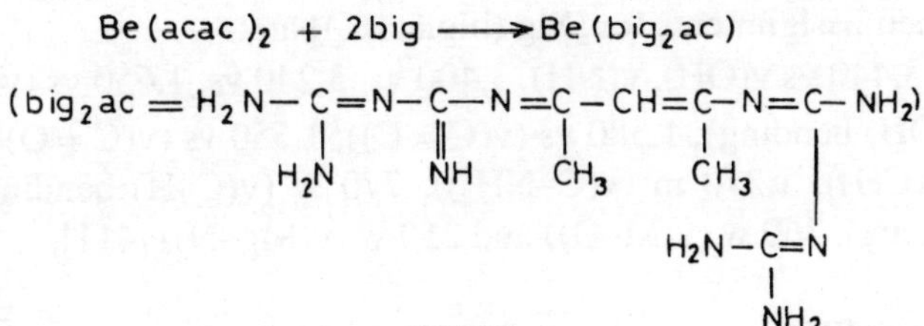

(XLIX)
Fig. 110

The reaction of bigH with $Be[CH(COCH_3)_2]_2$ gives $[Be(OH)big.H_2O]$. However, refluxing the same mixture yields $[Be(big)_2].2H_2O$ [411].

Action of the free base (bigH) on $[Be(acac)_2]$ in alcoholic medium yields a white precipitate of hydroxyaquamonobiguanido beryllium(II), $[Be(OH)(H_2O)(big)]$, best after prolonged refluxing white, crystalline $[Be(bigH)_2](OH)_2$ separate. The Be-N linkage is observed at 753 (s) and 769 (s) cm^{-1}—much lower than the ν(B-N) stretching vibrations (1,350-1,500 cm^{-1}). Such low frequencies my be ascribed to the highly electropositive character of beryllium or aluminum compared to boron [411].

(xxi) Aluminum(III)

$[Al(C_2H_5O)Cl(bigH)_2]Cl$ was prepared by the reaction between bigH and $AlCl_3$ in EtOH [412]. The Al-N bond is observed at 730 cm^{-1} and 770 cm^{-1}. The compound undergoes hydrolysis in aqueous media.

(xxii) Magnesium (II)

A white, crystalline Schiff base complex (m.p. 174-181°C) of magnesium was prepared by refluxing magnesium acetylacetonate with free bigH in dry ethanol (Fig. 111).

$$Mg(acac)_2 + bigH \longrightarrow Mg(big.acac_2).2H_2O$$

Fig. 111

Thermal analysis shows that water of crystallization is lost at 80-100°C [411].

The electronic spectrum of the solid shows three peaks at 310 nm, 262 nm, and 215 nm; in aqueous solution the peaks shift to 284 nm ($\varepsilon = 0.6$), 262 nm (sh) ($\varepsilon = 1.9$), and 232 nm ($\varepsilon = 0.15$). The first and last bands at 300 nm and 210 nm are due to $\pi \rightarrow \pi^*$ transitions located on the acacH and bigH chromophore, respectively [78, 413]. The bands due to the $\pi \rightarrow \pi^*$ transition of metal biguanides and metal acetylacetonates undergo blue and red shifts, respectively, as these complexes interact weakly among their respective π-orbitals. The completely new band appearing at 200 nm is due to the presence of a new chromophore in the Schiff base complexes [411]. Some important IR bands and their assignments for [Mg (big.acac$_2$)] are:

3,480 vs, 3,440 vs ν(OH), ν(NH), 3,400 vs, 3,240 vs, 1,650 vs (ν($\overline{N{-}C{-}N}$); 1,620 vs ν(OH) bending); 1,580 vs (ν(C = C)); 1,550 vs (ν(C = O)); 1,480 vs (ν(CO) + ν(CH)), 1,270 m (ν(C–NH$_2$)), 770 w (ν(C–H) bending), 730 w (ν(NH) bending), 360 w (ν(M–O)) and 250 w (ν(Mg–N)) [411].

(xxiii) Silicon(IV)

White, amorphous precipitates of compounds of the types $[Si(H_2O)_2(LH)_2]X_4.nH_2O$ (LH = bigH/AUH; X = Cl or NO_3; n = 1 or 2) and $[Si(C_6H_5)_2(LH)_2]I_2.2H_2O$ have been prepared by mixing an alcoholic solution of the appropriate ligand base with a benzene solution of tetrahalosilanes or diphenyldihalosilane at ice-salt temperature [134]. The free base in alcoholic media possesses sufficient π-electrons to interact with some of the bonding π-orbitals of the silicon atom so that a stable silicon heterocyclic compound is formed. Conductance measurements of the chloride salt show that the compound in aqueous medium dissociates into $[Si(H_2O)_4(LH)_2]^{4+}$ and X^- ions. These are classic examples of water-soluble cationic silicon(IV) complexes with stable Si–N bonds.

All the chelates and their ligands exhibit bands around 3,200-3,400 cm^{-1}. The sharp peak at 3,400 cm^{-1} due to the ν (N–H) stretch in biguanidinium ion shifts to 3,320 cm^{-1} in the silicon compound. This lowering of the frequency may be attributed to the Si–N bond, which has a profound effect on the N–H stretching frequency. Si–N bond formation increases the electron demand of the nitrogen and tends to prevent resonance in the guanidine residue. This in turn raises the C–N stretching frequency. Bands at 1,662 cm^{-1}, 1,673 cm^{-1}, and 1,680 cm^{-1} for the chelated compounds may be attributed to C–N stretching, and bands at 1,540 cm^{-1} and 1,605 cm^{-1} may be assigned to –NH$_2$ deformation [134].

(xxiv) Mercury(II)

In an alcoholic medium (palud) [HgCl] has been isolated by reaction of paludH and mercury(II) chloride [333].

(xxv) Lanthanides(III)

Albescu and Spacu [414-418] prepared substituted biguanide-paludH complexes of neodymium(III), samarium(III), europlum(III), gadolinium(III), terbium(III), dysprosium(III), holmium(III), and ytterbium(III). The degrees of covalency of these complexes were found to be greater than those of the corresponding $[Ln(H_2O)_6]^{3+}$ ion, showing that the ligand is bonded to the trivalent metal ion.

(xxvi) Uranium (VI)

Uranyl complexes of glucosyl-bigH, glucosyl(Me-bigH), glucosyl(Et-bigH), glucosyl(Me_2-bigH), glucosyl(Et_2-bigH) and glucosyl(Ph-bigH) have been prepared and characterized by elemental analysis and magnetic susceptibility measurements [245]. A diamagnetic uranyl complex, $[UO_2$(biguanidebenzohydroxamic acid)]$.2H_2O$, has also been prepared and characterized [203].

Complexes $(bigH_2)_2[UO_2F_4].H_2O$ and $(Ph\text{-}bigH_2)_2[UO_2F_4]$ were prepared by Chakraborty *et al* [419], who also isolated three new compounds by heating phenylbiguanide compounds. The TG curve of $(Ph\text{-}bigH_3)[UO_2F_4].2H_2O$ (Fig. 112) and $(bigH_2)_2[UO_2F_4].H_2O$ (Fig. 112) showed several distinct stages of decomposition [419].

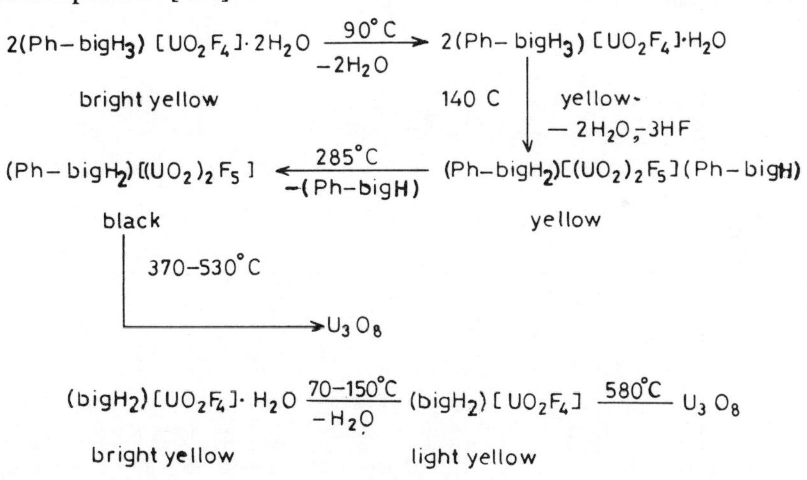

Fig. 112

(xxvii) Gold (III) and (I)

Yellow gold(III) biguanide complexes $[Au(bigH)_3]X_3$ (X = Cl, Br, or ClO_4) have been prepared by the addition of chloroauric acid solution to an alkaline solution of $bigH.H_2SO_4$. After digestion on a steam bath, the mass was filtered, and treatment of the residue with concentrated hydrochloric acid, hydrobromic acid, or perchloric acid yielded the corresponding salts [420]. Addition of chloroauric acid and hypophosphorous acid to an alkaline solution of bigH yielded [Au(big)(bigH)].H_2O, which, on treatment with concentrated acid hydrochloric acid, hydro bromic acid or perchloric acid yielded $[Au(bigH)_2]X$ (X = Cl, Br, or ClO_4) [420].

(xxviii) Zinc(II)

A number of zinc(II)-bigH/substituted-bigH complexes of the type $[ZnL_2]$ have been synthesized by reaction of sodium zincate and the ligand in a strongly alkaline medium. The compounds are white, silky crystals, readily hydrolyzed by water to yield a precipitate of zinc hydroxide. They were recrystallized from a hot solution of an alkali containing a small amount of the chelating ligand [1, 14, 194].

The stabilities of zinc(II) biguanides are less than those of the corresponding $[Zn\ en(bigH)_2]^{2+}$ salts [1]. $[Zn(bigH)_2]^{2+}$ is tetrahedral [112]. A TG study of $[Zn(big)_2]$ shows it to be stable to 225°C; at 355°C the compound begins to decompose, and at 590°C the pyrolytic residue becomes zinc oxide [208]. Syamal and Ghanekar reported zinc(II) complexes of trimethylenedibiguanide [112]. They could not isolate the cationic zinc(II) complex because $[Zn\ trime(big)_2]$ undergoes decomposition during the conversion process. The Zn-N bond is confirmed by the IR spectrum, which shows a weak bond at 500 cm^{-1} [112].

6

CONCLUSIONS

A. REACTIONS OF COORDINATED BIGUANIDES

Chromium(III) and Cobalt(III) biguanide base complexes react with mercury(II) chloride in aqueous media to form sparingly water soluble compounds of composition, $Cr(big.HgCl)_3$ and $Co(big.HgCl)_3.HgCl_2$, respectively – evidence that only one free amino ($-NH_2$) group in each biguanide molecule is bonded to the metal atom. On the other hand, free biguanide base reacts with mercury(II) chloride to form sparingly water soluble compound $C_2H_5N_5(HgCl)_2$ suggesting that both the amino ($-NH_2$) groups in the biguanide molecule are reactive. No such reaction occurs when the $[Co(en)_3]^{3+}$ cation reacts with mercury(II) chloride – evidence for the inertness of the bonded amino ($-NH_2$) group [122]. The formation of $[Co(big.AgOH)_3].nH_2O$ also supports this conclusion [122].

The Copper(II) N^1,N^1-diethylbiguanide base reacts with copper(II)-salicylaldehyde in methanol to form the 5, 5-diethylsalicylidenebiguanide copper (II) complex (Fig. 113) (XLX) [1]

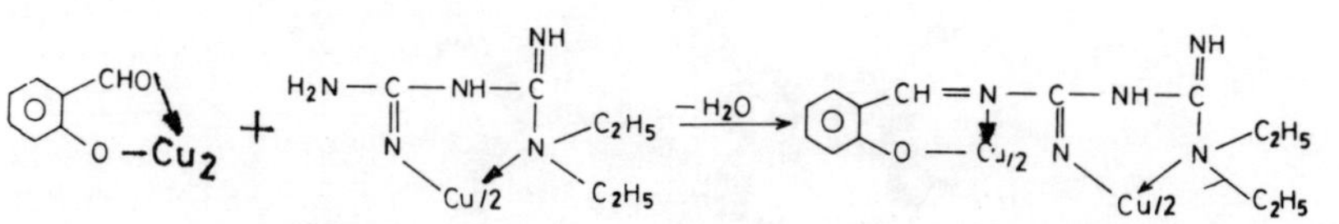

(XLX)

Fig. 113

Magnetic moment and magnetic susceptibility values over the range 80-310°K indicate the antiferromagnetic interaction of this complex (XLX). The absence of a band at 1540 cm^{-1} (phenoxide stretching before briding) but its presence at 1572-1580 cm^{-1} in ir spectra indicate phenoxide bridging in the complex. The shift of this band to a higher frequency is probably due to the increase in C—O bond strength for extended delocalisation of the π-system of the biguanide moiety. Considering the ir, and visible spectra and magnetic susceptibility measurements, Sen [212] discarded the structure (XLX) and proposed dimeric structure (Fig. 114) (XLXI) :

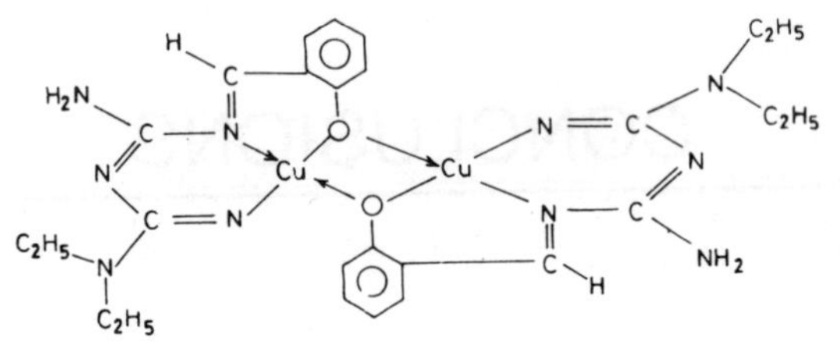

(XLXI)
Fig. 114

$[Ni(H_2O)_2(dichlorobigH)_2]Cl_2$ reacts with mercury(II) chloride and form $[Ni(dichloro\text{-}bigH.HgCl)_2]Cl_2.HgCl_2$. The chlorine atoms are substituted at the secondary amine nitrogen ($>C = NH$) atoms. Therefore the reaction with mercury(II) chloride occurs at the free amino ($-NH_2$) group.

The chlorination of square planar rose red $[Cu(bigH)_2]Cl_2.2H_2O$ in aqueous carbon tetrachloride medium gives blue octahedral trans-$[Cu(H_2O)_2(chloro\text{-}bigH)_2]Cl_2.2H_2O$. X-ray studies reveal that the substitution takes place on nitrogen atoms bonded to the copper(II) ion, and only one chlorine atom is substituted per biguanide molecule (Fig. 115) (XLXII). The metal-nitrogen bond distance in $[Cu(H_2O)_2(dichlorobigH)_2]Cl_2.2H_2O$ (Cu–N = 2.016 Å) is

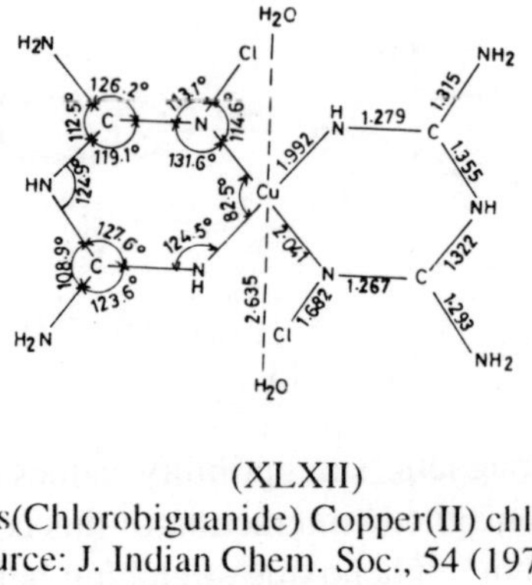

(XLXII)
Fig. 115 Bis(Chlorobiguanide) Copper(II) chloride dihydrate
Source: J. Indian Chem. Soc., 54 (1977) 133

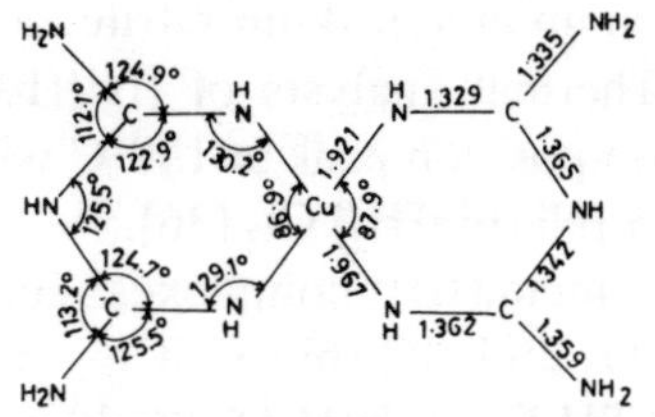

(XLXIII)

Fig. 116 Bis(biguanide) Copper(II) chloride dihydrate
Source: J. Indian Chem. Soc., 54 (1977) 133

greater than in [Cu(bigH)$_2$]Cl$_2$.2H$_2$O (Cu–N = 1.949 Å) (Fig. 116) (XLXIII) although a shortening of the ≡ C–NH$_2$ bond distance in the former compound is observed. Such enhanced metal-nitrogen bond distance and shortening of the carbon-nitrogen bond distance in the ≡ C–NH$_2$ group may be due to the drifting of the delocalised electron cloud towards the chlorine atom. The chlorine atom linked to biguanide weakens the σ and π donor capabilities of the nitrogen atoms. This is demonstrated by a comparion of the 10 Dq value of trans-[Cu(H$_2$O)$_2$(dichlorobigH)$_2$]Cl$_2$.2H$_2$O (16129 cm^{-1}) with that of [Cu(bigH)$_2$]Cl$_2$.2H$_2$O (18867.cm^{-1}). The two aqua molecules in [Cu(H$_2$O)$_2$(dichloro bigH)$_2$]Cl$_2$ are situated in the trans position at distances of 2.635 Å from the copper(II) ion. This is possible because the electron cloud may be localised to some extent toward the electronegative chlorine atom making room for the approaching water molecules. However, in [Cu(bigH)$_2$]Cl$_2$.2H$_2$O the π-electron cloud seems to prevent the water molecules from approaching the copper(II) ion in the complex [36].

When chlorine gas is passed through the suspension of yellow [Ni(bigH)$_2$]Cl$_2$ in carbon tetrachloride, a maroon red compound [Ni(bigH)$_2$Cl$_2$]Cl is formed, which, after addition of water molecules, changes to a greenish blue paramagnetic compound, [Ni(H$_2$O)$_2$(dichlorobigH)$_2$]Cl$_2$. A halo-substitution reaction mechanism is suggested [36] (Fig. 117):

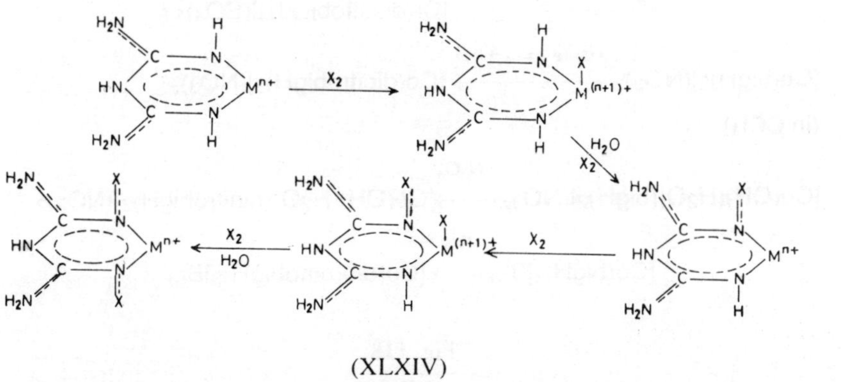

(XLXIV)
Fig. 117

On heating, $[Ni(H_2O)_2(dichlorobigH)_2]Cl_2$ explodes. It is reduced to $[Ni(bigH)_2]Cl_2$ by alkali, ammonia, sodium nitrite, sodium sulphite, sulphur dioxide and hydrazine. Thermal analysis of $[Ni(H_2O)_2(dichlorobigH)_2]Cl_2$ shows an exothermic decomposition peak at 150°C with loss of four chlorine atoms and the formation of $[Ni\ (bigH)_2]\ Cl_2$ [36].

Chlorination of other metal(III) complexes such as $[Co(bigH)_3]Cl_3$, $[Co(OH)(H_2O)(big)_2].3H_2O$ and $[Cr(bigH)_3]Cl_3$ leads to the formation of $[Co(dichlorobigH)_3]Cl_3.5H_2O$, $[Co(H_2O)_2(dichlorobigH)_2]Cl_3.3H_2O$ and $[Cr(dichlorobigH)_3]Cl_3.3H_2O$ respectively [36]. The thermal stability of halo-substituted metal complexes, follows the order Cu(II) > Co(III) > Cr(III) > Ni(II), but for non substituted metal complexes the order is Co(III) > Cr(III) > Cu(II) > Ni(II). [21, 36, 129]. Such low stability of halo-substituted metal complexes may be attributed to withdrawal of electron density from the chelate ring towards the electronegative halogen atom. The higher stability of halo-substituted complexes of copper(II) chelates may be due to mono substitution per biguanide ring. The N—X bond strength is influenced by the extent of electron delocalisation in the ring as it will extend to X atom through interaction of p_z-orbital of X with the ring π-system. The flow of electron density from the ring towards the halogen atom (X) is also confirmed by ir spectra. The asymmetric and symmetric bands for ν $\overline{(N—C—N)}$ (ring) group at ~ 1670 cm^{-1} and 1520 cm^{-1} in nonsubstituted complexes show considerable red shift in the corresponding halogeno-substituted complexes at 1630 cm^{-1} and 1450 cm^{-1} respectively. The band at ~760 cm^{-1} (N—H bending) present in simple complexes is absent in the corresponding halog-substituted complexes, indicating halosubstitution at the ring nitrogen atoms. Other reactions of coordinated biguanides are given below.

$$[Cr(big)_3]3H_2O\ (\text{in } CCl_4) \xrightarrow{SO_3\text{vapor, 0°C, 30 min}} [Cr(sulfobigH)_3](SO_4)_{1.5}$$

$$[Cr(sulfobigH)_3](SO_4)_{1.5} \xrightarrow{SO_3\text{, 1 hr.}} [Cr(disulfobigH)_3](SO_4)_{1.5}$$

$$[Cr(big)_3]3H_2O\ (\text{in } CCl_4) \xrightarrow{SO_3\text{vapor, 0°C, 4 hrs}^{*}} [Cr(disulfobigH)_3](SO_4)_{1.5}$$

$$[Co(bigH)_3](NO_3)_3\ (\text{in } CCl_4) \xrightarrow{N_2O_4\text{(gas), 4 hrs}} [Co(dinitrobigH)_3](NO_3)_3$$

$$[Co(OH)(H_2O)(bigH)_2](NO_3)_2 \xrightarrow{N_2O_4} [Co(OH)(H_2O)(dinitrobigH)_2](NO_3)_2$$

$$[Co(bigH)_3]Br_3 \xrightarrow{Br_2} [Co(dibromobigH)_3]Br_3$$

Fig. 118

The substituted complexes are readily hydrolysed by water with the production of free acid

$$>N-X+HOH \longrightarrow >N-H+HOX$$

($X = Cl, NO_2, SO_3H$)

Fig. 119

An increase in the Racah interelectronic repulsion parameter (B) and a decrease in 10 Dq values have been noticed in the halo-substituted complexes compared to the unsubstituted complexes (Table 23). The presence of a halogen atom in the ligand weakens σ and π-donor capacities of the nitrogen atom. The weaker σ-bonds give rise to an increase in the metal-nitrogen bond distances, which, in term, decreases the π-overlap between the ligand and metal orbitals. This observation is fully substantiated by comparing the X-ray structural data for halo-substituted and unsubstituted bis(biguanide)Copper(II)chloride dihydrate (Figs. 115, 116) (XLXII, XLXIII) [36].

Table 23 Electronic Spectral Data of Metal Complexes and Their Chloroderivatives [36]

Complex	*10 Dq (cm^{-1})*	*B (cm^{-1})*
$[Co(bigH)_3]Cl_3$	22500	450
$[Co(dichlorobigH)_3]Cl_3$	22020	520
$[Cr(bigH)_3]Cl_3$	19840	605
$[Cr(dichlorobigH)_3]Cl_3$	18160	750

B. SOME MORE INFRARED SPECTRA ON SOME SELECTED METAL COMPLEXES

Some important band positions of selected metal complexes and their assignments are listed in Table 24. All the metal complexes exhibit a broad intense band in the range 3200-3400 cm^{-1} assignable to the NH group stretching vibration. The OH stretching frequencies expected in this region for the hydrated complexes overlap with the NH vibration. The strong bands for C = N stretch, NH deformation, and OH deformation are observed in the range 1600-1775 cm^{-1}. In biguanide sulphate this band occurs at 1680 cm^{-1} but undergoes a red shift on complexation. Such lowering of the C = N stretching frequency on complexation may be attributed to a lowering of the C = N bond order and formation of a metal-nitrogen bond.

It is to be noted that the NH stretch, C = N stretch, and NH deformation frequencies are lower in uncharged metal complexes compared to the uncoordinated biguanide supporting nitrogen coordination of biguanide. In the metal complexes, the N–C–N stretch is observed at higher frequency in comparison to that of free biguanide. A new medium intensity, band appears around 1240-1310 cm^{-1} in all metal complexes. This may either be characteristic of the following fragment of the molecular structure (Fig. 120) (XLXV) or to the activation of some *ir* inactive band as a result of complexation [421]. Such a band is not present in the spectra of the free ligand or its salts.

(XLXV)

Fig. 120

In each series of metal complexes, the NH deformation frequencies are higher and the N–C–N stretching frequencies are lower for the complex salts than for the complex bases. Such observations may be interpreted in terms of the ligand field strength differences of the ligand in charged and uncharged complexes [139, 147]. Table 25 records some *ir* bands of en(bigH)$_2$.H_2SO_4, its metal complexes, and their assignments [310].

The studies reported here have established the strong field character of these fascinating ligands, which follow closely the l-amidino-O-alkylureas in their transition metal chemistry [14, 115, 423, 424]. A number of them have been used as analytical reagents [51, 422] and for the treatment of hypoglycaemia [425, 426], malaria [15, 428], filaria [429], and influenza [427]. The role of biguanides as germicidal, bacteriostatic, antihypoglycaemic, and anticarcinogenic agents is well documented [15, 16, 425-433]. Furthermore, they are starting materials for a variety of triazines [432] and heterocyclic compounds [434] of medicinal value. Additionally, either in the free state or in the complexed form, they take part in Schiff base-type reactions [211] giving rise to a wide range of products. They stabilize a number of unusual oxidation states of transition metals and yield coordination complexes with very large formation constants.

The ligands themselves have no aromatic character, but aromaticity develops as soon as they are bonded to either metal or nonmetal ions. Studies of the synthetic, structural, magnetic, spectroscopic, chromatographic, kinetic, mechanistic, and other behaviour of metal biguanides have contributed significantly to the richness of modern experimental and theoretical coordination chemistry. It is hoped that in the future additional extensive studies will be undertaken to explore further the donor properties of these ligands to an even greater extent than has been done to date.

Table 24 Infrared Spectral Data (cm^{-1}) for Free Biguanide, Biguanide Sulphate and Metal-Biguande Complexes

LH	*LH.H₂SO₄.H₂O*	*[VOL₂(H₂O)]*	*[CrL₃].H₂O*	*[Mn₂L₄(OH)₂]H₂O*	*[CoL₃].2H₂O*	*[NiL₂].2H₂O*	*[CuL₂].2H₂O*	*[ZnL₂]*	*Assignments*
(1)	*(2)*	*(3)*	*(4)*	*(5)*	*(6)*	*(7)*	*(8)*	(9)	(10)
3,450 s 3,200 sh 3,000 sh	3,400 s 3,100 s	3,330 s 3,150 s	3,450 s 3,350 s 3,225 s	3,350 s 3,200 s	3,400 m 3,310 m 3,275 w 3,190 w	3,400 s 3,350 vs 3,200 s	3,300 s	3,400 s 3,350 w 3,280 m 3,150 m	N-H stretch
1,680 s 1,640 s (sh)	1,710 s 1,680 m	1,660 s 1,630 s	1,670 s 1,640 m	1,620 s	1,675 m 1,625 m	1,620 s	1,600 vs	1,640 m	C = N stretch +N-H def. +O-H def.
1,520 s (sh) 1,470 (s) 1,420 s	1,560 s 1,470 m 1,405 w	1,560 vs 1,510 s	1,580 m 1,490 s 1,400 s	1,580 vs 1,480 s	1,590 s 1,560 m 1,450 m	1,590 s 1,510 s 1,480 s	1,580 vs 1,500 s(sh) 1,480 s	1,560 s 1,500 m	N-C-N stretch
		1,310 s 1,260 m	1,280 m	1,250 m	1,250 m	1,300 s	1,290 s	1,240 s	Ring vibration?
1,170 w 1,030 vw	1,205 w 1,120 vs* 1,080 vs* 100 m*	1,140 w 1,080 w	1,130 w 1,070 w	1,120 w	1230 w 1,050 w	1,120 w 1,070 w	1,220 w 1,160 w 1,120 w 1,060 m	1,050 w	C-N stretch

1,010 vw								
920 w(sh)	970 m*	975 m	970 w		950 w	950 w	940 w	
	880 w	750 m	760 w	780 w				
	830 w		720 w	690 m	690 w	790 m	820 s	
					(broad)	765 m	780 m	
						725 w	760 m	750 w
							700 m	

*Bands due to SO_4; LH = Free biguanide; $LH.H_2SO_4$ = Biguanide Sulphate

Source: J. Inorg. Nucl. Chem., 36 (1974) 3685

Table 25 IR Spectral Data of Free Biguanide (LH), Biguanide Sulphate ($LH.H_2SO_4.H_2O$), Ethylenedibiguanide Sulphate ($L'H.2H_2SO_4.5H_2O$) and Cobalt(III) Heterochelates

LH	*$LH.H_2SO_4.2H_2O$*	*$L'H.2H_2SO.5H_2O$*	*$[Co(acac)(L'H)]Cl_2.2H_2O$*	*$[Co(pic)(L'H)]Cl_2.H_2O$*	*$[Co(dipic)(L'H)]Cl.1.5H_2O$*	*$[Co(IDA)(L'H)]Cl.1.5H_2O$*	*Assignments*
3,450 s	3,400 s	3,380-3,300	3,400 m	3,405 m	3,395 m	3,390 s	N–H st.
			3,320 m	3,315 m	3,325 m	3,320 w	
3,200 sh			3,280 w	3,295 w	3,290 w	3,290 m	
3,000 sh	3,100 s	3,150-3,100	3,195 w	3,205 w	3,200 w	3,205 m	
1,680 s	1,710 s	1,696 s	1,690 vs	1,680 s	1,685 s	1,685 s	C = N st.
1,640 s(sh)	1,680 s	1,650 s	1,665 vs(br)	1,655 vs	1,650 vs	1,650 vs	+N–H def.
							+O–H def.
1,520 s(sh)	1,560 s	1,585	1,580 s	1,580 s	15,855 s	1,585 s	N–C–N st.
1,470 s	1,470 m		1,560 m	1,550 s	1,555 s	1,550 s	
1,420 s	1,405 w		1,455 m	1,460 s	1,465 s	1,460 m	
–	–	–	1,290	1,285	1,305	1,295	Ring vibration
1,170 w	1,205 w	1,375	1,225 w	1,230 w	1,220 w	1,225 w	C–N st.
	1,120 vs[a]	1,345	1,055 w	1,050 w	1,055 w	1,050 w	
1,030 vw	1,080 vs[a]						

1,010 vw						
920 w(sh)	970 m[a]	965	945 w	950 w	950 w	955 w
	880 w		690 w(br)	695 w(br)	690 w(br)	690 w(br)
	830 w					

All bands in cm^{-1}: Abbreviation: s = strong. vs = very strong, m = medium, sh = shoulder, w = weak, def. = deformation, st. = stretch, br = broad.
[a]Bands due to SO_4^{2-}. The literature IR band positions of picolinic acid (picH), 1690 cm^{-1}; acetylacetone (acacH). 1710. 1610 cm^{-1} (ref. 207(b)); biguanide hydrochloride (bigH.HCl), 1680, 1593, 1520 cm^{-1} (ref. 37).
Source : Polyhedron 11 (1992) 469.

Appendix

Appendix 1

Some Useful Physical and Chemical Constants

Atomic mass unit	amu	1.66057×10^{-27} kg
Avogadro number	N	6.022169×10^{23} mol^{-1}
Base, natural logarithms	e	2.71828
Bohr magneton	μ_B	9.274096×10^{-24} Am^2
Bohr radius	a_0	$5.2917715 \times 10^{-11}$ m
Boltzmann constant	k	1.38062×10^{-23} JK^{-1}
Debye	D	3.33564×10^{-30} Cm
Electronic charge	e	1.60210×10^{-19} C
Electronic rest mass	m_e	9.109558×10^{-31} kg
Faraday constant	F	9.648670×10^{4} C mol^{-1}
g (free electron)	g	2.002322
g value of *DPPH*	g_{DPPH}	2.0036
Gas constant	R	8.3143 JK^{-1} mol^{-1}
		8.2053×10^{-2} dm^3 atm K^{-1} mol^{-1}
Gravitational Constant ln	G	6.6720×10^{-11} Nm^2 . kg^{-2}
ln		2.3026
Permittivity of vacuum	ε_0	8.854185×10^{-12} C^2 m^{-1} J^{-1}
P_i	π	3.1415926536
Planck constant	h	6.6262×10^{-34} Js
	$\hbar = \frac{h}{2\pi}$	1.05459×10^{-34} Js
Proton mass	m_p	1.672614×10^{-27} kg
Rydberg constant	R	1.09737312×10^{7} m^{-1}
Speed of light	c	2.997925×10^{8} ms^{-1}

A — ampere; C — Coulomb; J — Joule; K — Kelvin; kg — Kilogram; kJ — Kilojoules; m — metre; mol — mole; s — second

Appendix 2

Useful Conversion Factors

Multiply	by	to obtain
Length		
cm	10^8	Å
cm	10^7	nm
cm	10^{10}	pm
Enegry		
kcal mol^{-1}	4.184	kJ mol^{-1}
eV	96.49	kJ mol^{-1}
Wave numbers (cm^{-1})	1.1962×10^{-2}	kJ mol^{-1}
kJ mol^{-1}	83.59	cm^{-1}
Dipole moments		
Debye	3.336×10^{-30}	Cm
Cm	0.300×10^{30}	D
Pressure		
atmosphere	1.013×10^5	Pa
Pascal	9.869×10^{-6}	atm

Appendix 3

Spectral Wavelength and the Colours

Range of the wavelength absorbed (nm)	*Colour of the radiation absorbed*	*Complementary colour observed*
400 - 450	Violet	Yellow-green
450 - 480	Blue	Yellow
480 - 490	Green-blue	Orange
490 - 500	Blue-green	Red
500 - 560	Green	Purple
560 - 575	Yellow-green	Violet
575 - 590	Yellow	Blue
590 - 600	Orange	Green-blue
600 - 650	Orange-red	Blue-green
650 - 800	Red	Blue

Appendix 4

A correlation table relating the symmetry species for several point groups with those for the point group O_h

O_h	T_d	D_{4h}	D_{2d}	C_{4v}	C_{2v}	D_{3d}
A_{1g}	A_1	A_{1g}	A_1	A_1	A_1	A_{1g}
A_{2g}	A_2	B_{1g}	B_1	B_1	A_2	A_{2g}
E_g	E	$A_{1g} + B_{1g}$	$A_1 + B_1$	$A_1 + B_1$	$A_1 + A_2$	E_g
T_{1g}	T_1	$A_{2g} + E_g$	$A_2 + E$	$A_2 + E$	$A_2 + B_1 + B_2$	$A_{2g} + E_g$
T_{2g}	T_2	$B_{2g} + E_g$	$B_2 + E$	$B_2 + E$	$A_1 + B_1 + B_2$	$A_{1g} + E_g$
A_{1u}	A_2	A_{1u}	B_1	A_2	A_2	A_{1u}
A_{2u}	A_1	B_{1u}	A_1	B_2	A_1	A_{2u}
E_u	E	$A_{1u} + B_{1u}$	$A_1 + B_1$	$A_2 + B_2$	$A_1 + A_2$	E_u
T_{1u}	T_2	$A_{2u} + E_u$	$B_2 + E$	$A_1 + E$	$A_1 + B_1 + B_2$	$A_{2u} + E_u$
T_{2u}	T_1	$B_{2u} + E_u$	$A_2 + E$	$B_1 + E$	$A_2 + B_1 + B_2$	$A_{1u} + E_u$

REFERENCES

1. P. Rây, *Chem. Rev.*, 61 (1961) 313.
2. R.L. Dutta and P. Rây, *J. Indian Chem. Soc.*, 36 (1959) 499.
3. P.F.B. Barnard, *J. Chem. Soc.*, A (1969) 2141.
4. W.A. Baker, Jr. and M. Daniels, *J. Inorg. Nucl. Chem.*, 25 (1963) 1194.
5. R.L. Dutta and A. Syamal, *Coord. Chem. Rev.*, 2 (1967) 441.
6. R.L. Dutta and R.K. Ray, *J. Indian Chem. Soc.*, 53 (1976) 507.
7. R.K. Ray, M.K. Bandyopadhyay and G.B. Kauffman, *Polyhedron*, 8 (1989) 757.
8. G. Wiedemann, *Ann.*, 68 (1848) 324.
9. M.J. Begley, P. Hubberstey and C.H.M. Moore, *J. Chem. Res.*, (S) (1985) 378; *J. Chem. Res.*, (M) (1985) 4001.
10. M.J. Begley, P. Hubberstey and C.H.M. Moore, *J. Chem. Res.*, (S) (1986) 120, 172; *J. Chem. Res.*, (M) (1986) 1033.
11. E. Baumann, *Ber.*, 7 (1874) 446.
12. E. Bamberger and W. Dickmann, *Ber.*, 25 (1892) 543.
13. H. Ley and F. Werner, *Ber.*, *46* (1913) 4040.
14. A. Syamal, *J. Sci. Ind. Res.*, 37 (1978) 661.
15. F.H.S. Curd and F.L. Rose, *J. Chem. Soc.*, (1946) 729.
16. G.D. Campbell, *Oral Hypoglycaemic Agents*, Academic Press, New York (1969).
17. L. Fabbrizzi, M. Micheloni, P. Paoletti and G. Schwarzenbach, *J. Am. Chem. Soc.*, 99 (1977) 5574.
18. G. Schwarzenbach and G. Anderegg, *Pharm. Acta, Helv.*, 38 (1963) 547.
19. A.K. De, N.N. Ghosh and P. Rây, *J. Indian Chem. Soc.*, 27 (1950) 493.
20. *Chem. Soc. Spec. Publication*, 25 (1971).
21. B. Das Sarma and P. Rây, *J. Indian Chem. Soc.*, 33 (1956) 841.
22. P.K. Das and A.K. De, *Proc. Convention of Chemists Allahabad, (India)*, (1972) 4.
23. M.L. Simms, J.L. Atwood and D.A. Zatko, *J. Chem. Soc., Chem. Commun.*, (1973) 46.
24. S.P. Ghosh and K.M. Prasad, *J. Less. Common Met.*, 36 (1974) 223.
25. D. Sen and C. Saha, *J. Chem. Soc., Dalton Trans.*, (1976) 776.
26. P. Rây and N.K. Dutt, *J. Indian Chem. Soc.*, 20 (1943) 81.
27. D. Banerjea and J.C. Bailar, Jr., *Transition Met. Chem.*, 10 (1985) 331.
28. D. Banerjea, *Transition Met. Chem.*, 13 (1988) 160.
29. T. Tada, Y. Kushi and H. Yoneda, *Bull. Chem. Soc. Japan*, 55 (1982) 1063.
30. M.R. Snow, *Acta Cryst.*, B30 (1974) 1850.

31. R.N. Banerjee, S. Gangopadhyay, A.K. Banerjee and D. Banerjea, *J. Coord. Chem.*, 12 (1983) 287.
32. A. Das, S. Gangopadhyay and D. Banerjea, *Transition Met. Chem.*, 12 (1987) 18.
33. R.N. Banerjee and D. Banerjea, *Indian J. Chem.*, 17A (1979) 246.
34. D. Banerjea, *Transition Met. Chem.*, 7 (1982) 22.
35. S. Guha and D. Banerjea, *Transition Met. Chem.*, 16 (1991) 264.
36. D. Sen and C. Saha, *J. Indian Chem. Soc.*, 54 (1977) 127.
37. C. Saha, D. Sen and S. Guha, *J. Chem. Soc.*, Dalton Trans., (1975) 1701.
38. M.K. De and R.L. Dutta, *J. Indian Chem. Soc.*, 52(1975) 67.
38(a) N.R. Singh, N.R. Singh and A.M. Singh, *J. Indian Chem. Soc.*, 72 (1995) 671.
39. M. K. De, D. Chatterjee and G.S. De, *J. Indian Chem. Soc.*, 67(1990) 420.
40. N.R. Singh, N.R. Singh and A.M. Singh, *J. Indian Chem. Soc.*, 70(1993) 665.
41. R.L. Dutta, R.K. Ray and G.B. Kauffman, *Coord. Chem. Rev.*, 28(1979) 23.
42. R.L. Dutta and R.K. Ray, *J. Indian Chem. Soc.*, 53(1976) 507.
43. R.K. Ray and G.B. Kauffman, *J. Chromatogr.*, 442(1988) 381; 504(1990) 464.
44. R.K. Ray, M.K. Bandyopadhyay and G.B. Kauffman, *J. Chromatogr.*, 469(1989) 383.
45. R.K. Ray and G.B. Kauffman, *Inorg. Chim. Acta*, 162(1989) 45.
46. R.K. Ray and G.B. Kauffman, *J. Indian Chem. Soc.*, 68 (1991) 291.
47. R.K. Ray and G.B. Kauffman, *Transition Met. Chem.*, 17 (1992) 141.
48. L. Fabbrizzi, M. Micheloni and P. Paoletti, *Inorg. Chem.*, 17 (1978) 494.
49. A.K. Ray, *Z. Anal. Chem.*, 156 (1957) 18.
50. P. Rây and J. Roy Choudhury, *J. Indian Chem. Soc.*, 18 (1941) 149.
51. F.J. Welcher, *Organic Analytical Reagents*, D. Van. Nostrand Co., Inc., New York, III (1967) 331.
52. A. Bhaduri and P. Rây, *J. Indian Chem. Soc.*, 22 (1945) 229.
53. F. Bosch Reig, J. Martinez Calatayud and M.C. Garcia Alvarez-Coque, *Afinidad*, 84 (1980) 394.
54. J.M. Martinez Calatayud, F. Bosch Reig and M.C. Garcia Alvarez-Coque, *Talanta, 29* (1982) 139.
55. K. Dey, R.K. Maity, J.K. Bhar, R.D. Banerjee, G.M. Sarkar, A. Malakar, S. Dutta and P. Banerjee, *Agents and Actions*, 11 (1981) 762.
56. K. Michelsen, *Acta Chem. Scand.*, 19 (1965) 1175.
57. G.R. Brubaker and L.E. Webb, *J. Am. Chem. Soc.,* 91 (1969) 7199.
58. K. Igi, T. Yasui, J. Hidaka and Y. Shimura, *Bull. Chem. Soc. Japan*, 44 (1971) 426.
59. N.C. Moucharafieh, P.G Eller, J.A. Bertrand and D.J. Royer, *Inorg. Chem.*, 17 (1978) 1220.
60. R.D. Peacock and B. Stewart, *Coord. Chem. Rev.*, 46 (1982) 129.
61. K. Igi, J. Hidaka and B. Douglas, *J. Coord. Chem.*, 7 (1978) 155.
62. H. Kawaguchi and B.E. Douglas, *J. Coord. Chem.*, 5 (1976) 115.
63. H. Kawaguchi, T. Ama and T. Yasui, paper presented at the *32nd Symposium on Coord. Chem.*, Nagasaki, Japan, Oct., (1982).
64. S. Tachiyashiki and H. Nakayama, paper presented at the *26th Symposium on Coord. Chem., Sapporo, Japan*, Aug., (1976).
65. H. Kawaguchi, M. Matsuki, T. Ama and T. Yasui, *Bull. Chem. Soc.* Japan, 59 (1986) 31.

66. S.R. Ernst and F.W. Cagle, Jr., *Acta Cryst.*, B33 (1977) 235.
67. S.R. Ernst, *Acta Cryst.*, B33 (1977) 237.
68. A.A. Pinkerton and D. Schwarzenbach, *J. Chem. Soc.*, Dalton Trans., (1978) 989.
69. D. Schwarzenbach and R. Schmelczer, *Acta Cryst.*, B34 (1978) 1827.
70. C.J. Brown, *J. Chem. Soc.*, A (1967) 60.
71. R. Handa and N.N. Saha, *Acta Cryst.*, B29 (1973) 559.
72. L. Coghi, M. Nardelli and G. Pelizzi, *Acta Cryst.*, B32 (1976) 842.
73. C.H.L. Kennard, G. Smith and E.J. O'Reilly, *Inorg. Chim. Acta*, 77 (1983) L113.
74. T. Tada, Y. Kushi and H. Yoneda, *Bull. Chem. Soc. Japan*, 54 (1981) 1538.
75. S.K. Hota, C.R. Saha and H. Pritzkov, *J. Coord. Chem.*, 13 (1984) 131.
76. N.R. Kunchur and M. Mathew, *J. Chem. Soc., Chem. Commun.*, (1966) 86.
77. M. Mathew and N.R. Kunchur, *Acta Cryst.*, B26 (1970) 2054.
78. D. Sen, *J. Chem. Soc.*, A (1969) 2900.
79. T.C. Creitz, R. Gsell and D.L. Wampler, *J. Chem. Soc., Chem. Commun.*, (1969) 1371.
80. A. Syamal, *Indian J. Chem. Educ.*, 7 (1980) 35.
81. J. R. Wasson and C. Trapp, *J. Phys. Chem.*, 73 (1969) 3763.
82. R.K. Ray, *Polyhedron*, 8 (1989) 7.
83. B. Rathke, *Ber.*, 12 (1879) 774.
84. R. Herth, *Wien. Akad. Ber.*, 80 (1881) 1078.
85. A. Ostrogovich, *Bull. Soc. Stiinte, Bucuresti*, 19 (1910) 64; *Chem. Zentr.*, II (1910) 1890.
86. P. Oxley and W.F. Short, *J. Chem. Soc.*, (1951) 1252.
87. K. Rachmann, *Ann.*, 376 (1910) 163.
88. K.H. Slotta and R. Tschesche, *Ber.*, 62 (1929) 1390; D. Karipides and W.C. Fernelius, *Inorg. Syn.*, 7 (1963) 56.
89. W. Cramer, *Ber.*, 34 (1901) 2594.
90. A.F. Reibenschuh, *Monatsh. Chem.*, 4 (1883) 388.
91. F. Emich, *Monatsh. Chem.*, 4 (1883) 395.
92. A. Smolka and A. Friedrich, *Monatsh. Chem.*, 9 (1888) 227.
93. F. Emich, *Monatsh. Chem.*, 12 (1891) 5.
94. S.P. Ghosh and A.K. Chatterjee, *J. Indian Chem. Soc.*, 32 (1955) 222.
95. A. Lumière, L. Lumière and F. Perrin, *Bull. Soc. Chim. France*, 33 (1905) 205.
96. P. Rây and N.N. Ghosh, *J. Indian Chem. Soc.*, 26 (1949) 144.
97. K. H. Slotta and R. Tschesche, *Ber.*, 62 (1929) 1398.
98. J.H. MacGregor, *U.S. Pat.* 2,448, 448; C.A., 43 (1949) 868.
99. N.R. Sengupta and P. Rây, *J. Indian Chem. Soc.*, 36 (1959) 373.
100. M.M. Ray, *Z. anorg. allg. Chem.*, 306 (1960) 133.
101. R.L. Dutta, *J. Indian Chem. Soc.*, 37 (1960) 32.
102. P. Rây and A.K. Chaudhury, *J. Indian Chem. Soc.*, 27 (1950) 651.
103. H. Schotte, H. Priewe and H. Roeschiesen, *z. physik. Chem.* 174 (1929) 119.
104. N. Saha and D. Mukherjee, *Polyhedron*, 3 (1984) 983.
105. N. Saha and D. Mukherjee, *Polyhedron*, 5 (1986) 1317.
106. S.R. Saha and A. Bagchi, *Indian J. Chem.*, 23A (1984) 579.
107. S.R. Saha and A. Bagchi, *Indian J. Chem.*, 24A (1985) 610.

108. S.R. Saha, A. Bagchi and A. Das, *J. Indian Chem. Soc.*, 67 (1990) 234.
109. E. Dittler, *Monatsh. Chem.*, 29 (1908) 645.
110. K. Chakraborty and P. Rây, *J. Indian Chem. Soc.*, 21 (1944) 41.
111. P. Rây, *Inorg. Syn.*, 6 (1960) 75; *Nature*, 151 (1943) 643.
112. A. Syamal and V.D. Ghanekar, *J. Inorg. Nucl. Chem.*, 40 (1978) 1606.
113. I.F. Kurzer and E.D. Pitchfork, *Fortschr. Chem. Forsch.*, 10 (1968) 375.
114. K. Sugino and M. Ogawa, *Rev. Phys. Chem. Japan*, 13 (1939) 58.
115. F.H.S. Curd, D.G. Davey and D.N. Richardson, *J. Chem. Soc.*, (1949) 1732.
116. N. Kundu and P. Rây, *J. Indian Chem. Soc.*, 29 (1952) 811.
117. R. Passerini, *Boll. sci., fac. chim. ind. Univ. Bologna*, 9 (1951) 27; C.A., 45 (1951) 10208.
118. G. Pelizzari, *Gazz. chim. ital.*, 53 (1923) 384.
119. R. Banerjee, A. Bhattacharya, P.K. Das and A.K. Chakraburtty, *J. Chem. Soc.*, Dalton Trans., (1988) 1557.
120. R. Banerjee, A. Bhattacharya and A.K. Chakraburtty, *Transition Met. Chem.*, 14 (1989) 55.
121. N.K. Misba and G.B. Mitra, *Z. Krist.*, 119 (1963) 155.
122. P. Rây and S.K. Siddhanta, *J. Indian Chem. Soc.*, 18 (1941) 298.
123. G. Schwarzenbach and D. Schwarzenbach, *J. Indian Chem. Soc.*, 54 (1977) 23.
124. B. Das Sarma, *J. Indian Chem. Soc.*, 29 (1952) 217.
125. A.K. Ray and B. Das Sarma, *Z. anorg. allg. Chem.*, 305 (1960) 207.
126. N.R. Sengupta and P. Rây, *J. Indian Chem. Soc.*, 37 (1960) 303.
127. R.L. Dutta and N.R. Sengupta, *J. Indian Chem. Soc.*, 38 (1961) 741.
128. S.D. Nandi, *Tetrahedron*, 28 (1972) 845.
129. R.L. Dutta, *J. Indian Chem. Soc.*, 37 (1960) 499.
130. J.C. Gage, *J. Chem. Soc.*, (1949) 221.
131. B. Chakraborty, P.K. Das and A.K. Sil, *Inorg. Chim. Acta*, 30 (1978) 149.
132. G. Hung-Yiu Liu and D.J. MacDonald, *J. Inorg. Nucl. Chem.*, 31 (1969) 3233.
133. A. Das, A. Bagchi and S.R. Saha, *Indian J. Chem.*, 29A (1990) 361.
134. A. Maitra and D. Sen, *Inorg. Nucl. Chem. Lett.*, 8 (1972) 793.
135. E. Cristurean, P. Spacu and C. Gheorghiu, *J. Less Common Metals*, 21 (1970) 15.
136. P. Spacu and C. Gheorghiu, *Z. anorg. allg. Chem.*, 364 (1969) 331.
137. A. Chakraborty and S. Basu, *J. Inorg. Nucl. Chem.*, 17 (1961) 55.
138. S.D. Nandi and P. Banerjee, *Z. anorg. allg. Chem.*, 406 (1974) 124.
139. P.V. Babykutty, C.P. Prabhakaran, R. Anantaraman and C.G.R. Nair, *J. Inorg. Nucl. Chem.*, 36 (1974) 3685.
140. M.M. Ray and P. Rây, *J. Indian Chem. Soc.*, 36 (1959) 849.
141. R.L. Dutta and D. De, *J. Indian Chem. Soc.*, 46 (1969) 169.
142. A. Chakraborty and S. Basu, *Nature*, 185 (1960) 681.
143. D.J. MacDonald, *Inorg. Chem.*, 6 (1967) 2269.
144. P. Spacu, C. Gheorghiu and A. Nicolaescu, *Inorg. Chim. Acta*, 2 (1968) 413.
145. P. Spacu and C. Gheorghiu, *Z. anorg. allg. Chem.*, 351 (1967) 201.
146. R.L. Dutta and S. Sarkar, *J. Indian Chem. Soc.*, 44 (1967) 832.
147. R.H. Skabo and P.W. Smith, *Aust. J. Chem.*, 22 (1969) 659.
148. C.R. Saha, *Indian J. Chem.*, 15A (1977) 794.

149. R.S. Banerjee and S. Basu, *J. Inorg. Nucl. Chem.*, 27 (1965) 2452.
150. R.L. Dutta and T. Ray, *J. Indian Chem. Soc.*, 58 (1981) 709.
151. D. Banerjea and S. Dutta Chaudhuri, *J. Inorg. Nucl. Chem.*, 30 (1968) 871.
152. F. Woldbye, *Acta. Chem. Scand.*, 12 (1958) 1079.
153. R.L. Dutta and A.M. Singh, *J. Indian Chem. Soc.*, 52 (1975) 1063.
154. D. Sen, *Int. Conf. Coord. Chem.*, 9 (1966) 236.
155. P. Spacu and C. Gheorghiu, *J. Less Common Metals*, 15 (1968) 331.
156. A.I.P. Sinha and S.P. Ghosh, *Sci. Cult.*, 27 (1961) 312.
157. S.P. Ghosh and A.I.P. Sinha, *J. Inorg. Nucl. Chem.*, 26 (1964) 1703.
158. S.P. Ghosh and A.I.P. Sinha, *J. Indian Chem. Soc.*, 41 (1964) 330.
159. C. Gheorghiu, *Z. anorg. allg. Chem.*, 365 (1969) 91.
160. R.S. Banerjee and S. Basu, *J. Inorg. Nucl. Chem.*, 27 (1965) 353.
161. B. Chatterjee and A. Syamal, *J. Indian Chem. Soc.*, 47 (1970) 1021.
162. A. Syamal, *Indian J. Pure & Appl. Phys.*, 21 (1983) 130.
163. P. Spacu and C. Gheorghiu, *J. Less Common Metals*, 18 (1969) 117.
164. L. Coghi, M. Lanfranchi, G. Pelizzi and P. Tarascono, *Transition Met. Chem.*, 3 (1978) 69.
165. A.N. Mitra and D. Sen, *J. Inorg. Nucl. Chem.*, 34 (1972) 3643.
166. C.R. Saha, *Indian J. Chem.*, 16A (1978) 862.
167. R.S. Banerjee and S. Basu, *J. Inorg. Nucl. Chem.*, 27 (1965) 359.
168. J. Bera and D. Sen, *Indian J. Chem.*, 14A (1976) 880.
169. P. Rây and N. Kundu, *J. Indian Chem. Soc.*, 29 (1952) 811.
170. P. Rây *J. Indian Chem. Soc.*, 32 (1955) 141.
171. G.D. Diana, E.S. Zalay and R.A. Cutler, Jr., *J. Org. Chem.*, 30 (1965) 298.
172. L. Tschugaeff, *Ber.*, 40 (1907) 1975.
173. H. Ley and F. Müller, *Ber.*, 40 (1907) 2950.
174. W. Traube and G. Glaubitt, *Ber.*, 63 (1930) 2094.
175. P. Rây and H. Saha, *J. Indian Chem. Soc.*, 14 (1937) 670.
176. W.E. Swartz, Jr. and R.A. Alfonso, *J. Electron Spectrosc. and Relat. Phenom.*, 4 (1974) 351.
177. A. Syamal, *Chem. Educ.*, 4 (1987) 33; 5 (1988) 26.
178. R.L. Dutta, *Inorganic Chemistry*, New Book Stall, Calcutta (1980) 212; (1996) 351.
179. K.N. Goswami and S.K. Dutta, Z. Krist., 120 (1964) 399.
180. K.N. Goswami and S.K. Dutta, *Indian J. Phys.*, 37 (1963) 604.
181. B.L. Holian and R.E. Marsh, *Acta Cryst.*, B26 (1970) 1049.
182. D.L. Ward, C.N. Caughlan and G.D. Smith, *Acta Cryst.*, B27 (1971) 1541.
183. L. Coghi and G. Pelizzi, *Acta Cryst.*, B31 (1975) 131.
184. L. Coghi, A. Mangia, M. Nardelli and G. Pelizzi, *Ricerca Scient.*, 39 (1969) 438.
185. L. Coghi, A. Mangia, M. Nardelli, G. Pelizzi and L. Sozzi, *J. Chem. Soc., Chem. Commun.*, (1968) 1475.
186. G.D. Andreetti, L. Coghi, M. Nardelli and P. Sagarabotto, *J. Cryst. Molecular Struct.*, 1 (1971) 147.
187. N.R. Kunchur, *Nature*, 217 (1968) 539.
188. K. Marjatta, L. Elsa and N. Reino, *Suom. Kemistilehti*, B43 (1970) 254; C.A., 73 (1970) 102975w.
189. K.N. Goswami, *Z. Krist.*, 122 (1965) 473.

190. S.K. Ray and S. C. Chakraborty, *Indian J. Phys.*, 40 (1966) 267.
191. Index to X-ray Powder Data File, ASTM, Philadelphia, Publication 48 L (1962) 75
192. C. Gheorghiu, M. Constantinescu and O.V. Pocol, *Rev. roum. Chim.*, 19 (1974) 1009.
193. C. Gheorghiu and M. Bichir, *An. Univ. Bucuresti. Ser. Stiint. Natur. Chim.*, 22 (1973) 129.
194. N.R. Sengupta, *Z. anorg. allg. Chem.*, 315 (1962) 220.
195. E.D. Weinberg, R. Chernin and J.H. Billman, *J. Am. Pharm. Assn.*, 49 (1960) 441.
196. W.O. Foye, R.L. O'Laughlin and R.N. Duvall, *J. Pharm. Sci.*, 50 (1961) 641.
197. C. Gheorghiu and E. Cristurean, *An. Univ. Bucuresti. Ser. Stiint. Natur. Chim.*, 15 (1966) 61.
198. M. Yamada and K. Odo, *Yuki Gosei, Kagaki Kyokai. Shi.*, 21 (1963) 946.
199. C. Gheorghiu and L. Antonescu, *An. Univ. Bucuresti. Ser. Stiint. Natur. Chim.*, 13 (1964) 151.
200. S.P. Ghosh and A.K. Banerjee, *J. Indian Chem. Soc.*, 41 (1964) 275.
201. C.W. Catron, D.R. Lorentz and J.R. Wasson, *Inorg. Nucl. Chem. Lett.*, 12 (1976) 385.
202. R.L. Dutta and S. Lahiri, *Z. anorg. allg. Chem.*, 306 (1960) 116.
203. S.N. Poddar, M.M. Ray and A.K. Sarkar, *Sci. Cult.*, 20 (1963) 308.
204. A. Anichini, L. Fabbrizzi, P. Paoletti and R.M. Clay, *J. Chem. Soc., Chem. Commun.*, (1977) 244.
205. M. Ciampolini, P. Paoletti and L. Sacconi, *J. Chem. Soc.*, (1960) 4553.
206. D. Banerjea, *Indian J. Chem.*, 26A (1987) 543; *J. Indian Chem. Soc.*, 67 (1990) 286.
207. S. Guha and D. Banerjea, *Transition Met. Chem.*, 16 (1991) 441.
208. P.V. Babykutty, P. Indrasenan, R. Anantaraman and C.G.R. Nair, *Thermochim. Acta*, 8 (1974) 271.
209. N.N. Ghosh and A. Bhattacharya, *J. Indian Chem. Soc.*, 48 (1971) 689.
210. S.R. Saha and A. Das, *Indian J. Chem.*, 32A (1993) 802.
211. S.N. Poddar and A.K. Sarkar, *J. Indian Chem. Soc.*, 45 (1968) 11, 979.
212. J.L. Bera, P. Pramanik and D. Sen, *J. Chem. Soc.*, Dalton Trans., (1975) 2436.
213. B. Datta, K. Singh and D. Sen, *Electrochim. Acta, 25* (1980) 533.
214. B. Datta and K. Singh, *Indian J. Chem.*, 20A (1981) 926.
215. S.N. Poddar, N.G. Poddar and S.M. Bhattacharya, *Indian J. Chem.*, 13 (1975) 393.
216. D.A. Doornbos, *Pharm. Weekbl. Ned.*, 103 (1968) 1213.
217. J.R. Siefker and R.L. Wence, *Proc. Indiana Acad. Sci.*, 75 (1965) 104.
218. D. Banerjea, *Z. anorg. allg. Chem.*, 344 (1966) 72.
219. R.L. Dutta and R.K. Ray, *J. Indian Chem. Soc.*, 51 (1974) 187.
220. R.L. Dutta and R.K. Ray, *J. Indian Chem. Soc.*, 52 (1975) 579.
221. P. Rây, *Inorg. Syn.*, 7 (1963) 6.
222. P. Rây and S.K. Siddhanta, *J. Indian Chem. Soc.*, 20 (1943) 250.
223. P. Rây and R.K. Dutt, *J. Indian Chem. Soc.*, 25 (1948) 563.
224. J.R. Wasson and P. Zacharopoulos, *J. Inorg. Nucl. Chem.*, 31 (1969) 3881.
225. W. Schneider and P. Baccini, *Helv. Chim. Acta*, 52 (1969) 1955.
226. W. Schneider and A.von. Zelewsky, *Helv. Chim. Acta,* 48 (1965) 1529.

227. P. Spacu, C. Gheorghiu, O. Constantinescu and I. Pascaru, *Z. physik. Chem.*, 65 (1969) 248.
228. P. Spacu, C. Gheorghiu, O. Constantinescu and I. Pascaru, *Revue roum. Chim.*, 17 (1971) 667.
229. P. Spacu, C. Gheorghiu, M. Constantinescu, I. Pascaru and O. Constantinescu, *Z. anorg. allg. Chem.*, 384 (1971) 184.
230. A. Syamal, *J. Indian Chem. Soc.*, 64 (1987) 719.
231. A. Syamal, *Indian J. Chem.*, 24A (1985) 946.
232. R.K. Ray and G.B. Kauffman, *Inorg. Chim. Acta*, 174 (1990) 257.
233. R.K. Ray and G.B. Kauffman, *Inorg. Chim. Acta*, 174 (1990) 237.
234. W.R. Rice, H.J. Stoklosa, J.R. Wasson and A. Syamal, *Proc. Int. Conf. Coord. Chem.*, 14 (1972).
235. C.R. Saha and S.K. Hota, *Indian J. Chem.*, 25A (1986) 340.
236. D. Sen, A.K. Panda and B. Banerjee, *Indian J. Chem.*, 21A (1982) 741.
237. R.L. Dutta and S.P. Banerjee, *Indian J. Phys.*, (July 1976), (Commemoration Vol., Pt. 1 (1976) 301.
238. D. Banerjea, *Proc. 25th Int. Conf. Coord. Chem.*, Nanjing, China (1987) 831; *Pure & Appl. Chem.*, 60 (1988) 1337; *J. Indian Chem. Soc.*, 67 (1990) 269.
239. T. Pramanik, A. Das and D. Banerjea, *Transition Met. Chem.*, 16 (1991) 332.
240. R.L. Dutta and R.K. Ray, *Indian J. Chem.*, 52 (1975) 387.
241. R.L. Dutta and R.K. Ray, *J. Inorg. Nucl. Chem.*, 41 (1979) 755.
242. R.L. Dutta, S.P. Banerjee and D. De, *J. Indian Chem. Soc.*, 52 (1975) 582.
243. R.L. Dutta, D. De and A. Syamal, *J. Indian Chem. Soc.*, 44 (1967) 353.
244. R.L. Dutta, D. De and A. Syamal, *J. Indian Chem. Soc.*, 45 (1968) 663.
245. S.N. Poddar, *Sci. Cult.*, 29 (1963) 219.
246. E. Buluggiu, G. Dascola, D.C. Giori and V. Varacca, *Phys. Status Solidi*, 45B (1971) 217.
247. J. R. Siefker and L.J. Jardine, *Proc. Indiana Acad. Sci.*, 77 (1977) 176.
248. D.J. MacDonald, *J. Inorg. Nucl. Chem.*, 29 (1967) 1271.
249. D.J. MacDonald, *J. Inorg. Nucl. Chem.*, 30 (1968) 1971.
250. P. Spacu and C. Gheorghiu, *Z. Chemie, Lpz., 8* (1968) 184.
251. P. Spacu and I. Albescu, *Rev. Chim. Acad. Rep. Populare Romaine*, 7 (1962) 179.
252. P. Spacu and I. Albescu, *Acad. Rep. Populare Romaine Studi Leratari Chim.*, 8 (1960) 85, 91.
253. P. Capitan and P. Salinas, Bull. Soc. Quim. Peru, 33 (1967) 1.
254. C. Gheorghiu, A. Nicolaescu and A.L. Pausescu, *An. Univ. Bucuresti. Ser. Stiint. Natur. Chim.*, 21 (1972) 105.
255. B. Dutta and D. Sen, *Indian J. Chem.*, 19A (1980) 669.
256. D. Sen, *Sci. Cult.*, 28 (1962) 292.
257. P. Spacu, C. Gheorghiu and E. Cristurean, *Z. Chem.*, 7 (1967) 28.
258. D.J. Gulliver and W. Levason, *Coord. Chem. Rev.*, 46 (1982) 1.
259. D.K. Nag and S. Guha, *Indian J. Phys.*, 45 (1971) 139.
260. D. Banerjea and P. Banerjee, *Indian J. Chem.*, 8 (1970) 373; *Z. anorg. allg. Chem.*, 393 (1972) 295.
261. P. Capitan and P. Salinas, *Rev. Soc. Quim. Mex.*, 11 (1967) 178.
262. C. Gheorghiu and E. Cristurean, *An. Univ. Bucuresti. Ser. Stiint. Natur. Chim.*, 15 (1966) 153.

263. P. Rây and S.P. Ghosh, *J. Indian Chem. Soc.*, 20 (1943) 327.
264. E. Cristurean, *An. Univ. Bucuresti. Ser. Stiint. Natur. Chim.*, 22 (1973) 137.
265. E. Cristurean, *An. Univ. Bucuresti. Ser. Stiint. Natur. Chim.*, 18 (1969) 147.
266. P. Spacu, C. Gheorghiu and E. Cristurean, *Revue roum, Chim.*, 9 (1964) 265.
267. P. Spacu, *An. Univ. Bucuresti. Ser. Stiint. Natur. Chim.*, 12 (1963) 79.
268. C. Gheorghiu and E. Cristruréan, *An. Univ. Bucuresti. Ser. Stiint. Natur. Chim.*, 15 (1966) 131.
269. M.M. Ray, *J. Indian Chem. Soc.*, 36 (1959) 860.
270. W.D. Closson and H.B. Gray, *J. Am. Chem. Soc.*, 85 (1963) 290.
271. D. Sen, *J. Indian Chem. Soc.*, 51 (1974) 183.
272. P. Spacu, C. Gheorghiu and N. Vladescu, *Z. Chem.*, 6 (1966) 188.
273. P. Spacu, C. Gheorghiu and N. Vladescu, *Revue roum. Chem.*, 12 (1969) 269.
274. C. Gheorghiu and N. Vladescu, *An. Univ. Bucuresti. Ser. Stiint. Natur. Chim.*, 15 (1966) 101.
275. S.S. Gupta and R. Kaushal, *J. Indian Chem. Soc.*, 51 (1974) 649.
276. P. Rây, *Inorg. Syn.*, 6 (1960) 71.
277. R.L. Dutta, A. Bhattacharya and M.K. De, *J. Indian Chem. Soc.*, 52 (1975) 371.
278. A.E. Martell and M. Calvin, *Chemistry of the Metal Chelate Compounds*, Prentice-Hall, Englewood Cliffs, NJ (1952) 518.
279. A. Syamal, *Indian J. Phys.*, 42 (1968) 120.
280. D. Banerjea, *J. Indian Chem. Soc.*, 38 (1961) 642.
281. D. Banerjea and B. Chakraborty, *J. Inorg. Nucl. Chem.*, 26 (1964) 1233.
282. D. Banerjea and B. Chattopadhyay, *J. Inorg. Nucl. Chem.*, 36 (1974) 2351.
283. B. Chakraborty and A.K. Sil, *Indian J. Chem.*, 15A (1977) 502.
284. P. Rây and S.P. Ghosh, *J. Indian Chem. Soc.*, 19 (1942) 1; 20 (1943) 291.
285. R.N. Banerjee and D. Banerjea, *J. Indian Chem. Soc.*, 55 (1978) 969.
286. B. Chakraborty, S. Modak and B. Nandi, *Indian J. Chem.*, 21A (1982) 480.
287. S.F. Mason, *Inorg. Chim. Acta.*, Rev., 2 (1968) 89.
288. P. Spacu, *Rev. Chim. Acad. Rep. Populare Romaine*, 7 (1962) 561.
289. R.K. Ray and G.B. Kauffman, *J. Chromatogr.*, 675 (1994) 271.
290. R.K. Ray and M.K. Bandyopadhyay, paper presented at the *12th World Chromatography Conference*, Capri, Italy (1989).
291. B. Datta, K. Singh and D. Sen, *Indian J. Chem.*, 21A (1982) 958.
292. R.L. Dutta and I. Mallik, *J. Indian Chem. Soc.*, 44 (1967) 866.
293. R.K. Ray and G.B. Kauffman, *Synth. React. Inorg. Metal. Org. Chem.*, 21 (1991) 1417.
294. A. Syamal, P.K. Mondal and S.K. Verma, *Transition Met. Chem.*, 3 (1978) 288.
295. R.L. Dutta and A. Syamal, *J. Indian Chem. Soc.*, 45 (1968) 1014.
296. B.C. Bag and A. Syamal, *J. Indian Chem. Soc.*, 47 (1970) 475.
297. R.L. Dutta and K.K. Bhattacharya, *J. Indian Chem. Soc.*, 45 (1968) 1091.
298. R.L. Dutta and S. Sarkar, *J. Indian Chem. Soc.*, 44 (1967) 853.
299. S.P. Ghosh and P. Bhattacharjee, *Indian J. Chem.*, 14A (1976) 501
300. S.K. Chatterjee, P.K. Das and S.K. Bose, *J. Indian Chem. Soc.*, 54 (1977) 1099.
301. R.L. Dutta, S. Sarkar and K.K. Bhattacharya, *J. Indian Chem. Soc.*, 50 (1973) 235.

302. R.L. Dutta and S. Sarkar, *Sci. Culture*, 30 (1964) 549.
303. A. Syamal and P.K. Mondal, *Transition Met. Chem.*, 3 (1978) 288, 292; *J. Indian Chem. Soc.*, 57 (1980) 100; *Transition Met. Chem.*, 4 (1979) 348.
304. A. Syamal and P.K. Mondal, *J. Inorg. Nucl. Chem.*, 43 (1981) 609.
305. N.K Ray and C.R. Saha, J. Inorg. Nucl. Chem., 42 (1980) 37.
306. R.L. Dutta and A. Bhattacharya, *J. Indian Chem. Soc.*, 52 (1975) 1002.
307. R.L. Dutta and A. Bhattacharya, *J. Indian Chem. Soc.*, 57 (1980) 429.
308. R.K. Ray and G.B. Kauffman, *Polyhedron*, 8 (1989) 1015.
309. R.K. Ray, *Polyhedron*, 9 (1990) 2245.
310. R.K. Ray, *Polyhedron*, 19 (1992) 469.
311. R.L. Dutta and S. Sarkar, *J. Indian Chem. Soc.*, 44 (1967) 842.
312. R.K. Ray and G.B. Kauffman, *Polyhedron*, 13 (1994) 2591.
313. B. Chakraborty, S. Modak and P.K. Das, *Indian J. Chem.*, 20A (1981) 97.
314. B. Chakraborty, S. Modak, P.K. Das and A.K. Sil, *J. Inorg. Nucl. Chem.*, 43 (1981) 3253.
315. B. Chakraborty, U.K. Kar and I. Ghosh, Indian J. Chem., 24A (1985) 103.
316. B. Chakraborty and A.K. Sil, *J. Indian Chem. Soc.*, 54 (1977) 84.
317. B. Chakraborty, I. Ghosh and U.K. Kar, *Indian J. Chem*, 25A (1986) 725.
318. B. Chakraborty and I. Ghosh, *Indian J. Chem.*, 24A (1985) 582.
319. R.L. Dutta, *J. Indian Chem. Soc.*, 44 (1967) 863.
320. A.K. Gangopadhyay and G.S. De, *Indian J. Chem.*, 26A (1987) 824; 27A (1988) 778; *Transition Met. Chem.*, 14 (1989) 241.
321. B. Chakraborty and A.K. Sil, *J. Indian Chem. Soc.*, 55 (1978) 211.
322. B. Chakraborty and A.K. Sil, *Indian J. Chem.*, 16A (1978) 312.
323. H.C. Dhar, *Ph.D. Thesis*, Burdwan University (India), (1988).
324. M. Pal and G.S. De, *Indian J. Chem.*, 14A (1976) 763.
325. M. Pal, *Ph.D. Thesis*, Burdwan University (India) (1977).
326. W. Kruse and H. Haube, *J. Am. Chem. Soc.*, 83 (1961) 1280.
327. P.M. Brown and G.M. Harris, *Inorg. Chem.*, 7 (1968) 1872.
328. D.N. Hague and J. Halpern, *Inorg. Chem.*, 6 (1967) 2059.
329. N.M. Samus and T.S. Lukyanets, *Zh. Neorg. Khim.*, 20 (1975) 1010.
330. B. Chakraborty and P.K. Das, *Indian J. Chem.*, 16A (1978) 583.
331. S.P. Ghosh, J.N. Gupta and P. Bhattacharjee, *Proc. 3rd Symp. Coord. Chem.*, Debrecen, Hungary, 1 (1970) 201.
332. S.P. Ghosh and J.N. Gupta, *J. Indian Chem. Soc.*, 33 (1956) 193.
333. R. Tsuchiya, A. Uehara, K. Otsuka and E. Kyuno, *Chem. Lett.*, (1974) 833.
334. Y. Ihara, A. Uehara, R. Tsuchiya and E. Kyuno, *Bull. Chem. Soc. Japan*, 51 (1978) 2578.
335. C. Gheorghiu and A. Nicolaescu, *An. Univ. Bucuresti. Ser. Stiint. Natur. Chim.*, 14 (1965) 143.
336. A. Syamal and P.K. Mondal, *J. Indian Chem. Soc.*, 55 (1978) 515.
337. P. Rây and H. Saha, *J. Indian Chem. Soc.*, 14 (1937) 675, 680.
338. J.C. Brassington, R.J.P. Williams and P.E. Wright, *J. Chem. Soc., Chem. Commun.*, (1975) 338.
339. D. Banerjea and S. Sengupta, *Z. anorg. allg. Chem.*, 397 (1973) 215; 424 (1976) 93.
340. D. Banerjea, *Transition Met. Chem.*, 12 (1987) 97.
341. P Rây and H. Saha, *J. Indian Chem. Soc.*, 15 (1938) 633.

342. A. Syamal, *Z. Naturf.*, 29B (1974) 492.
343. R.L. Dutta and T. Ray, *J. Inorg. Nucl. Chem.*, 43 (1981) 1985.
344. A.K. Gangopadhyay and G.S. De, *Current Science*, 56 (1987) 1214.
345. A.K. Gangopadhyay and G.S. De, *Indian J. Chem.*, 28A (1989) 899.
346. C. Gheorghiu and C. Guran, *An. Univ. Bucuresti. Ser. Stiint. Natur. Chim.*, 21 (1972) 83.
347. A. Swinarski, H. Zawadzki and P. Przemyslau, *Roczn. Chem.*, 49 (1975) 1239, 1249, 1987.
348. A. Swinarski and H. Zawadki, *Roczn. Chem.*, 50 (1976) 2021.
349. D. Banerjea and P. Choudhuri, *J. Inorg. Nucl. Chem.*, 32 (1970) 2697.
350. C. Gheorghiu, *Revue roum. Chim.*, 13 (1968) 1181.
351. A.I.P. Sinha and S.P. Ghosh, *J. Indian Chem. Soc.*, 38 (1961) 179.
352. S.P. Ghosh and A.I.P. Sinha, *J. Indian Chem. Soc.*, 40 (1963) 249.
353. C. Gheorghiu and L. Burdea, *An. Univ. Bucuresti. Ser. Stiint. Natur. Chim.*, 21 (1972) 53.
354. P. Spacu, *An. Univ. Bucuresti. Ser. Stiint. Natur. Chim.*, 13 (1974) 179.
355. S.P. Ghosh and P. Bhattacharjee, *J. Inorg. Nucl. Chem.*, 32 (1970) 573.
356. J. Lewis, R.S. Nyholm and P.W. Smith, *J. Chem. Soc.*, (1961) 4590.
357. B. Chakraborty, P.K. Das and A.K. Sil, *Proc. Convention of Chemists (India), Inorg.*, (1977) 69.
358. S.P. Ghosh and R.K. Prasad, *J. Indian Chem. Soc.*, 64 (1987) 765.
359. B. Chakraborty and A.K. Sil, *Inorg. Chim. Acta*, 24 (1977) 105.
360. S.P. Ghosh and U. Prasad, *J. Less Common Metals*, 44 (1976) 345.
361. D. Sen, *Sci. Cult.*, (Calcutta) 28 (1962) 84.
362. P. Spacu and C. Gheorghiu, *Revue roum. Chim.*, 12 (1967) 1459.
363. B. Chakraborty, S. Adhikari and U.K. Kar, *Indian J. Chem.*, 29A (1990) 903.
364. P. Rây and K. Chakraborty, *J. Indian Chem. Soc.*, 21 (1944) 49.
365. R. Banerjee, K. Das, A. Das and S. Dasgupta, *Inorg. Chem.*, 28 (1989) 585.
366. S.N. Poddar and N.G. Poddar, *J. Indian Chem. Soc.*, 47 (1970) 39.
367. D. Sen, *J. Chem. Soc.*, A, (1969) 1304.
368. C.R. Saha, *J. Inorg. Nucl. Chem.*, 38 (1976) 1635.
369. W. Levason and M.D. Spicer, *Coord. Chem. Rev.*, 76 (1987) 45.
370. A.K. Banerjee and S.P. Ghosh, *J. Indian Chem. Soc.*, 53 (1976) 719.
371. R.M. Sanyal, B.H. Ansari, P.C. Srivastava, B.K. Banerjee and A.K. Chakraborty, *Fert. Technol.*, 16 (1979) 36.
372. H.N. Pσ, *Coord. Chem. Rev.*, 20 (1976) 171.
373. J.A. MeMillan, *Chem. Rev.*, 62 (1962) 65.
374. D. Sen and P. Rây, *J. Indian Chem. Soc.*, 30 (1953) 519.
375. D. Sen, N.N. Ghosh and P. Rây, *J. Indian Chem. Soc.*, 27 (1950) 619.
376. R. Banerjee, R. Das and S. Mukhopadhyay, *J. Chem. Soc.*, Dalton Trans., (1992) 1317.
377. B. Chakraborty, *J. Inorg. Nucl. Chem.*, 41 (1979) 757.
378. D.A. Zatko and J.W. Prather, *J. Electron Spectrosc. Relat. Phenom.*, 2 (1973) 191.
379. D. Karweik, N. Winograd, D.G. Davis and K.M. Kadish, *J. Am. Chem. Soc.*, 96 (1974) 591.
380. B.A. Sok, *Diss. Abstr. Int. B*, 30 (1970) 3539.
381. J.D. Miller, *J. Chem. Soc. A*, (1968) 1778.

382. S. Dasgupta, A. Das and R. Banerjee, *Transition Met. Chem.*, 14 (1989) 19.
383. R. Banerjee, A. Das and S. Dasgupta, *J. Chem. Soc., Dalton Trans.*, (1989) 1645.
384. R. Banerjee, *Proc. Indian Acad. Sci. (Chem. Sci.)* 106 (1994) 655.

385. A. Maitra, *J. Indian Chem. Soc.*, 50 (1973) 361.
386. C. Gheorghiu and L. Antonescu, *An. Univ. Bucuresti. Ser. Stiint. Natur. Chim.*, 20 (1971) 113.
387. P. Spacu and L. Antonescu, *Revue roum. Chim.*, 17 (1972) 1851.
388. C. Gheorghiu and L. Antonescu, *An. Univ. Bucuresti. Ser. Stiint. Natur. Chim.*, 20 (1971) 89.
389. L. Antonescu, *An. Univ. Bucuresti. Ser. Stiint. Natur. Chim.*, 21 (1972) 135.
390. L. Antonescu and G. Kalmutchi, *Rev. Chim. (Bucuresti)*, 24 (1974) 923; 25 (1974) 68.
391. P. Spacu, C. Gheorghiu, M. Constantinescu and L. Antonescu, *J. Less Common Metals,* 44 (1976) 161.
392. M.M. Ray, K. De and S.N. Poddar, *Indian J. Chem.*, 3 (1965) 228.
393. B. Chakraborty, *J. Inorg. Nucl. Chem.*, 41 (1979) 1211.
394. M.M. Ray and P. Rây, *J. Indian Chem. Soc.*, 35 (1958) 595.
395. M.M. Ray, J.N. Adhya, D. Biswas and S.N. Poddar, *Aust. J. Chem.*, 19 (1966) 881.
396. N.G. Poddar and S.N. Poddar, *Aust. J. Chem.*, 21 (1968) 801.
397. K. Dey, K.C. Ray, S.N. Poddar and N.G. Poddar, *Indian J. Chem.*, 14A (1976) 205.
398. M.M. Ray and P. Rây, *Sci. and Cult. (Calcutta)*, 25 (1959) 384.
399. M.M. Ray, *J. Inorg. Nucl. Chem.*, 27 (1965) 2193.
400. S. Roychowdhury, M.K. Sipani and B. Sur, *J. Indian Chem. Soc.*, 65 (1988) 718.
401. N. Ta and D. Sen, *J. Inorg. Nucl. Chem.*, 43 (1981) 209.
402. N. Ta, P. Pramanik and D. Sen, *J. Indian Chem. Soc.*, 51 (1974) 374.
403. M.M. Ray, *Sci. Cult.*, 30 (1964) 190; *C.A.*, 61 (1964) 12932.
404. C. Gheorghiu, *An. Univ. Bucuresti. Ser. Stiint. Natur. Chim.*, 17 (1968) 31.
405. C. Gheorghiu, *An. Univ. Bucuresti. Ser. Stiint. Natur. Chim.*, 18 (1969) 31.
406. S. Lahiri and V.K. Anand, *Inorg. Chem.*, 20 (1981) 2789.
407. F. Bosch Reig, J. Martinez Calatayud and M.C. Garcia Alvarez-Coque, *Afinidad*, 37 (1980) 137.
408. G. Cohn, *J. prakt. Chem.*, 84 (1911) 394.
409. D. Sen, *Sci. Cult.*, 30 (1964) 190.
410. A.N. Mitra and D. Sen, *Indian J. Chem.*, 12 (1974) 183.
411. C.R. Saha, *J. Inorg. Nucl. Chem.*, 42 (1980) 159.
412. S.D. Nandi and P. Banerjee, *Z. Naturforsch.*, 29B (1974) 347.
413. R.H. Holm and F.A. Cotton, *J. Am. Chem. Soc.*, 80 (1958) 5658.
414. I. Albescu, *Revue roum. Chim.*, 17 (1972) 1729.
415. I. Albescu, *Revue roum. Chim.*, 17 (1972) 1867.
416. I. Albescu, *Revue roum. Chim.*, 18 (1973) 599.
417. I. Albescu, *Revue roum. Chim.*, 18 (1973) 829.
418. P. Spacu and I. Albescu, *Revue roum. Chim.*, 16 (1971) 1503.

419. M.C. Chakraborty and M. Bandyopadhyay, *J. Inorg. Nucl. Chem.*, 33 (1971) 2565.
420. D. Sen, *Sci. Cult.*, 27 (1961) 548.
421. K. Nakamoto, *In Coordination Chemistry* (Edited by A.E. Martell) Van Nostrand, New York, Vol. 1 (1970).
422. A.K. Majumdar, *J. Indian Chem. Soc.*, 20 (1943) 289.
423. R. Nongmathem, R. Nongmathem, J. Akham, S. Langpoklakpam and M. Akoijam, *J. Indian, Chem. Soc.*, 72 (1995) 665.
424. N.R. Singh, N.R. Singh, A.M. Singh, *J. Indian, Chem. Soc.*, 72 (1995) 671.
425. I.C. Bygbjerg, *Eur. J. Clin. Pharmacol.*, 28 (1985) 287.
426. C.J. Bailey, *Int. Congr., Symp. Ser. R. Soc. Med.*, 79 (1985) 17.
427. A. Pradalier and H. Casassus, *Diabete*, 22 (1974) 203.
428. M. Angelo, D. Ortwine, D. Worth, L.M. Werbel and J.W. McCall, *J. Med. Chem.*, 26 (1983) 1258.
429. S. Kashiwagi, *Fukuoka-Igaku-Zasshi*, 65 (1974) 157.
430. J. Z. Byozkowski and C.W. Porter, *Gen. Pharmacol.*, 14 (1983) 615.
431. B.K. Paul, S.S. Banerjee-Chaudhury, A.N. Bose and U.P. Basu, *Indian J. Chem.*, B14 (1976) 887.
432. V.G. Baspalov and V.A. Aleksandrov, *Byull. Eksp., Biol., Med.*, 100 (1985) 73.
433. R.C. Mehrotra, in *Comprehensive Coordination Chemistry*, G. Wilkinson, R.D. Gillard and J.A. McCleverty (Eds.), Pergamon, Oxford, New York, Vol. 2 (1987) 269-291.
434. R. Reinhard and V. Henri, *J. Org. Chem.*, 46 (1981) 3011.

AUTHORS INDEX

I

K

L

M

N

P

R

S

T

SUBJECT INDEX